Digital Integrated Circuits

Design-for-Test Using Simulink® and Stateflow®

Digital
Integrated
Circuits

Design-for-Test Using Simulink® and Stateflow®

Evgeni Perelroyzen

CRC Press
Taylor & Francis Group
Boca Raton   London   New York

CRC Press is an imprint of the
Taylor & Francis Group, an informa business

CRC Press
Taylor & Francis Group
6000 Broken Sound Parkway NW, Suite 300
Boca Raton, FL 33487-2742

International Standard Book Number-10: 0-8493-3057-2 (Hardcover)
International Standard Book Number-13: 978-0-8493-3057-5 (Hardcover)

Library of Congress Cataloging-in-Publication Data

Perelroyzen, Evgeni.
    Digital integrated circuits : design-for-test using Simulink and Stateflow / Evgeni Perelroyzen.
        p. cm.
    Includes index.
    ISBN-13: 978-0-8493-3057-5 (alk. paper)
    ISBN-10: 0-8493-3057-2 (alk. paper)
    1. Digital integrated circuits--Testing. 2. Digital integrated circuits--Design and construction. 3. SIMULINK. 4. MATLAB. I. Title. II. Title: Design-for-test using Simulink and Stateflow.

TK7874.P445 2007
621.3815--dc22   2006021811

Visit the Taylor & Francis Web site at
http://www.taylorandfrancis.com

and the CRC Press Web site at
http://www.crcpress.com

To Natalija Silova

Thy glass will show thee how thy beauties wear,
Thy dial how thy precious minutes waste,
The vacant leaves thy mind's imprint will bear,
And of this book this learning mayst thou taste.

William Shakespeare, LXXVII

Contents

Preface

This book's main objective is the construction of Simulink®* models for digital project test benches in the field of design-for-test. This book is a part of the novel tendency (described in the Introduction) of integrating the MATLAB® system (specifically its two components — Simulink and Stateflow®) into the process of modern digital design. The first part of the book describes the major tools used by the author for design-for-test: Simulink and Stateflow.

Chapter 1, which deals with Simulink, describes the process of Simulink model building, as well as the main blocks of libraries in Simulink, which the author uses for Simulink model building in the design-for-test field. Chapter 2, Stateflow, describes the process of finite state machine model building as Stateflow diagrams.

The second part of the book discusses the Simulink model building for some of the latest design-for-test fields: fault modeling and simulation for combinational circuits and sequential circuits (Chapter 3); Simulink model building for combinational controllability and observability and sequential controllability and observability computation (Chapter 4); the automatic test pattern generation (ATPG) process and Simulink model building for the D-algorithm and PODEM-algorithm (Chapter 5); logical determinant theory, digital circuit dynamics, and model building for timing verification (Chapter 6); and models for built-in self-test (BIST) architecture, scan cell operations, functional testing, diagnostic testing, and JTAG interface models (Chapter 7).

The author wishes to express his deepest and sincere gratitude to Ms. Koudloh Marina for her invaluable assistance in the book preparation.

* Simulink, Stateflow, and MATLAB are registered trademarks of The MathWorks, Inc. For product information, please contact: The MathWorks, Inc., 3 Apple Hill Drive, Natick, MA 01760-2098 USA. Tel: 508-647-7000. Fax: 508-647-7001. E-mail: info@mathworks.com. Web: www.mathworks.com.

Introduction

I.1 MATLAB Integration to Modern Digital Design Flow

At present, a novel trend in digital design is being developed: the integration of the MATLAB® system. It is, first and foremost, associated with digital signal processing.

On the intellectual properties (IP) side, the majority of design systems start operating by using the MathWorks' industry-leading products, Simulink® and MATLAB. Simulink provides an environment for IP building-blocks-based circuit engineering design as well as a project simulation environment and "what-if" analysis. Field programmable gate array (FPGA) vendors Xilinx, Altera, and Lattice and electronic design automation (EDA) vendor Synplicity, for instance, use Simulink to develop synthesis tools and IP blocks for the implementation of IP-based design flow. Thus, Synplicity's digital signal processing (DSP) Synthesis Tool starts its operation with model generation in the Simulink environment, using the blocks of its pertinent Simulink library. Further on, to obtain the final design results, the company's own register transfer level (RTL) Synthesis Technology is applied to the model [1]. We can cite a number of detailed examples.

I.1.1 MATLAB and Simulink Accelerate Implementation of DSP Designs on FPGAs: The result of collaboration between The MathWorks and Xilinx

The MathWorks deals with accelerated engineering design, using the MATLAB system and its Simulink and Stateflow® components, which are intended for the development and analysis of algorithms and for the system-level design of DSP applications. The MathWorks strategy for DSP applications is the following:

1. Acceleration of innovations in the communications and multimedia market
2. Achievement of key positions in the development of DSP algorithms
3. Acquisition of system-level design solutions, using Simulink and MATLAB
4. Support of integration between top-down design and project synthesis (implementation) tools

The advantages of The MathWorks and Xilinx collaboration are a seamless route from Simulink and MATLAB to highly efficient implementation tools of Xilinx DSP applications and less time taken by FPGA design and verification.

The disadvantages of the existing design are:

1. The gap between system-level design (using MATLAB and C language) and the implementation tools (EDA tools for the implementation of FPGA, DSP, and embedded software (SW) tool-based devices tailored for the implementation of controlling software and DSP applications)
2. Ambiguous specifications
3. Design problems that are revealed at advanced stages
4. High risk of design faults

The gap between system-level design and its implementation tools results in a lack of integration in the design team: they break up into two independent groups (supervisory logic group (DSP group) and analog and mixed signal group, each having its own tools). This results in the impossibility of simulating the interaction between the components devised and the impossibility of testing the entire designed system and makes redesign an extremely high-cost task.

The Simulink application as a system-level design tool brings into existence a common tool for all design groups, enables the simulation of interactions between the system's designed components and the testing of the entire designed system's behavior, and eliminates the necessity of redesign. The Simulink design environment provides top-down design, clear-cut specifications, design and test for the entire system, simulation of interactions between the designed components, early detection of design faults, and reduced time to market [2].

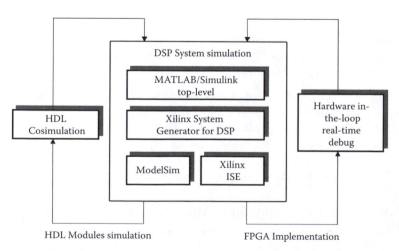

FIGURE I.1
Xilinx System Generator for Simulink.

Within the approach frame, Xilinx has developed the Xilinx System Generator™ for Simulink (Figure I.1). Before the Xilinx System Generator was implemented, design using Simulink was characterized by discordance between system architects (who create the block diagrams of Simulink models) and FPGA developers (who redesign the system architects' models, implementing the hardware description languages (HDLs: VHSIC HDL, Verilog HDL). Following close interaction and numerous iterations, they finally achieved concord, and only after that did FPGA developers synthesize the model and implement it physically.

With the advent of the Xilinx System Generator, system architects and FGPA developers could collaborate in the Simulink environment, using two major components as follows: the library of dedicated blocks, Xilinx Blockset, for model building in the Simulink environment; and the HDL generator that uses the Xilinx optimized IP algorithms and generates the synthesized HDL code with consequent physical implementation of the project as FPGA, using the Smart-IP™ tool [3, 4].

The work of Rabey and Chandrakasan [5] exemplifies computer implementation of the advanced base-band processor for a wireless modem based on its high-level description in the Simulink environment. This high-level description is converted into an engineering implementation in the "chip-in-a-day"-type design environment. This tool controls synthesis from the behavioral to the gate level. The project's complete physical implementation and verification takes about 24 hours. Such an approach is also efficient when high-level signal-processing functions are realized at rapid-prototyping platforms such as FPGAs. The Xilinx System Generator uses such models as filters, modulators, and correlators (described in MathWorks Simulink environment) directly in FPGA module.

By applying MATLAB and Simulink in combination with the Xilinx System Generator for DSP applications, one can simulate and verify the DSP algorithms at a certain platform (target hardware platform), without the necessity of leaving the Simulink environment. In this process, the design flow consists of the following stages (Figure I.2) [6]:

1. The DSP designer designs and verifies a hardware model, using the industry-standard tools of The MathWorks together with the Xilinx System Generator for DSP applications.

2. After the button with Xilinx System Generator icon is pressed, the Xilinx System Generator generates the HDL circuit representation.

3. The process is synthesized, using the ISE design tools, and the generated bitstream is employed in FPGA programming. Thus, the DSP designer can generate the FPGA bitstream directly in the Simulink–Xilinx System Generator environment.

XtremeDSP™ Development Kit II, developed by Nallatech in collaboration with Xilinx, is the platform to be used for the development of high-performance

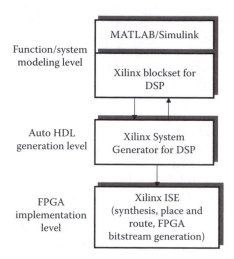

FIGURE 1.2
Design flow using Xilinx System Generator for Simulink.

signal processing systems [7]. The combination of XtremeDSP, Simulink, and the Xilinx System Generator gives the FPGA and DSP engineers the chance for fast and efficient hardware implementation of complete systems. Such an approach reduces the time to market for products that can reconfigure and modify the system without changing its hardware composition, as compared to the conventional HDL-based design.

When using the Xilinx System Generator, the model is created as follows: The Xilinx System Generator icon is activated in a Simulink environment at the model upper level. Then the window that opened in the Xilinx System Generator provides various options such as a FPGA type, housing type, synthetic tool type, and a location for the generated VHDL code. All these ensure prompt system design out of standard Xilinx System Generator blocks. If a designer wants to use the VHDL code, he or she can place it into the corresponding "black box" block.

Combinations of various Xilinx System Generator blocks and subsystems allow hierarchical system design placement of such blocks in the Simulink environment and support a bottom-up design approach, and as it progresses, parts of the model are tested and verified using various Simulink blocks such as Scope, Graph, Display, and so on. The MATLAB functions for pseudo three-dimensional visualization, image plotting, and calculations can be used for verification.

The software model is verified using the hardware Simulink cosimulation library block, created by the Xilinx System Generator, that controls the design flow. A compilation object is chosen in such a block for hardware cosimulation and so that XtremeDSP is the hardware target environment. The hardware Simulink cosimulation library block can be efficiently programmed by the FPGA bitstream, using the Xilinx Synthesis Tool (XST). The block makes possible certain operations such as data transfer and clocking, device configuration,

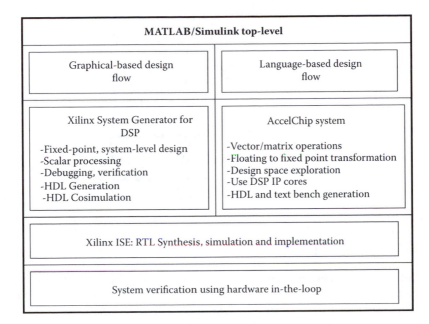

MATLAB/Simulink top-level	
Graphical-based design flow	Language-based design flow
Xilinx System Generator for DSP -Fixed-point, system-level design -Scalar processing -Debugging, verification -HDL Generation -HDL Cosimulation	**AccelChip system** -Vector/matrix operations -Floating to fixed point transformation -Design space exploration -Use DSP IP cores -HDL and text bench generation
Xilinx ISE: RTL Synthesis, simulation and implementation	
System verification using hardware in-the-loop	

FIGURE I.3
System Generator/Accel Chip Interface.

and so on. The simulation block library is created like any other Simulink library. The simulation is conducted in Simulink environment, and its results are transferred from the FPGA to XtremeDSP.

Two kinds of electronic designers are known: those who deal with language forms and those who deal with block diagrams. The Xilinx System Generator for DSP is a proficient tool for DSP project development by system architects and hardware designers who use the visual design environment (Simulink and Xilinx Block Set). The AccelChip Inc. system, called the AccelChip DSP Synthesis Tool, has been specifically developed for algorithm developers and DSP architects who prefer operating with language forms. Having started from the implementation of MATLAB system M files to the development of input stimuli and algorithm assessments, the project can later be exported to the Xilinx System Generator (Figure I.3), and the project optimization can be based on Xilinx FPGAs [8].

I.1.2 Altera DSP Builder: The Result of Collaboration of The MathWorks and Altera

DSP system design in an Altera programmable logic device (PLD) environment demands certain design tools for the implementation of high-level algorithms and project descriptions using HDLs. Altera DSP Builder® is precisely the development tool for DSP applications, and it connects the Simulink (the development facility for system-level DSP applications) to Altera's Quartus® system.

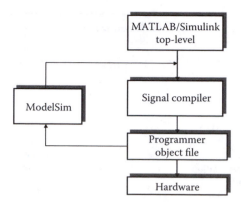

FIGURE I.4
Design flow using Altera DSP Builder.

Altera DSP Builder shortens the design stage, assisting designers in representing the project hardware implementation in the algorithm-friendly development environment. In this design, the MATLAB functions, the Simulink and Altera DSP Builder blocks, and the Altera Intellectual Property (IP) MegaCore functions are used, providing designers of systems, algorithms, and hardware with a single common development platform and ensuring seamless design. The design flow by means of the Altera DSP Builder is shown in Figure I.4 [9].

I.1.3 The New Tendency in Design-for-Test: Direct Link between Simulink and Hardware Description Language (HDL) Simulators to Enable Rapid Creation and Verification of System-Level Test Benches

The tendency of MATLAB to be integrated into numerical design has influenced the design-for-test domain. The collaboration of The MathWorks and Mentor Graphics® resulted in the Link for ModelSim® for use with MATLAB and Simulink [10] system, intended for VHDL project simulation and verification using Simulink. Integration of ModelSim, Simulink, and Simulink Blocksets ensures a powerful and efficient environment for simulation and cosimulation in electronic design automation (EDA).

The bond between ModelSim and Simulink permits integration of the hardware component models of ModelSim and alternative Simulink-located general model constituents into a single model. Two possible scenarios are available: (a) the only VHDL Cosimulation block is placed in the Simulink model environment or (b) the Simulink model represents the set of VHDL Cosimulation blocks, each of them being the hardware component model.

The standard operation order for the integration of VHDL Cosimulation blocks into the Simulink project with one or more hardware components

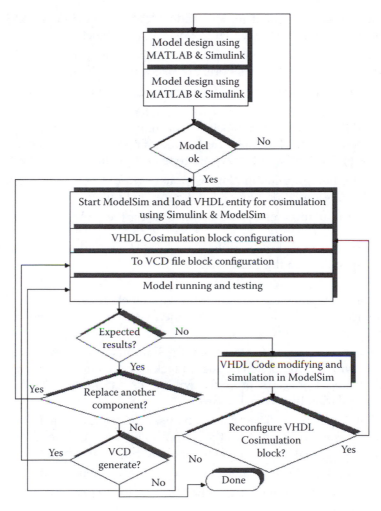

FIGURE I.5
The algorithm for VHDL Cosimulation blocks integration in the Simulink project.

consists of the following steps (illustrated by Figure I.5):

1. A user develops an applied model in Simulink. The model can contain, for instance, one or more components the user intends to describe using the VHDL.

2. The model is started up and tested in the Simulink environment.

3. If the model's operation is faulty, steps 1 and 2 are repeated to eliminate the faults.

4. The ModelSim is employed to simulate the VHDL-encoded project's discrete components.

5. The model component representation on VHDL is integrated into the Simulink model as a VHDL Cosimulation block.

6. The VHDL Cosimulation block is further configured: the block parameters describing its ports are fixed in the dialog window.

7. The model is restarted and tested in the Simulink environment.

8. If the model operates in a faulty manner, the VHDL code is modified, the VHDL Cosimulation block is resimulated in the ModelSim environment, and the block is reconfigured, if necessary. In other words, steps 6, 7, 8 (or 7 and 8 only) are repeated to eliminate the faults.

9. If any other Simulink model component should be replaced on the VHDL Cosimulation block, step 4 is carried out again.

10. The cosimulation results are verified, using the To VCD File block.

If a user wants to include Simulink in the EDA technological chain, he or she must speculate on the following.

1. Is the user planning to start from the development of the VHDL application, using ModelSim, and then test the results systemically in the Simulink environment?

2. Is the user planning to start from the Simulink environment system level with "hardware component black boxes," and, if this system-level model operates properly, will he or she replace the VHDL Cosimulation block black boxes?

3. What Simulink blocksets should be used? (Generally, the following blocksets are of interest for the EDA applications: Communications Blockset, Signal Processing Blockset, and Simulink Fixed Point.)

4. Will the VHDL Cosimulation blocks represent the model's subsystems?

5. What will the simulation time be? Will scaling be necessary?

6. Will the VCD file be generated?

After answering these questions, a user can employ Simulink to create the simulation environment. Figure I.6 shows the Simulink model that includes the Link for ModelSim block. The Link for ModelSim VHDL Cosimulation block bridges the Simulink and ModelSim simulation environments. The block is a hardware-component VHDL model in the Simulink environment. Simulink generates input signals for the block and reads its output signals after its simulation in ModelSim. Exchange of signals between the two systems is made in conformity with the time steps, determined by the Simulink simulation time.

After a Simulink model is developed, it should be started up and fully tested before its implementation in the Simulink Link for ModelSim blocks environment. The testing results are preserved and used later for the comparison of the starting Simulink model and the Simulink model that contains the Link for ModelSim blocks.

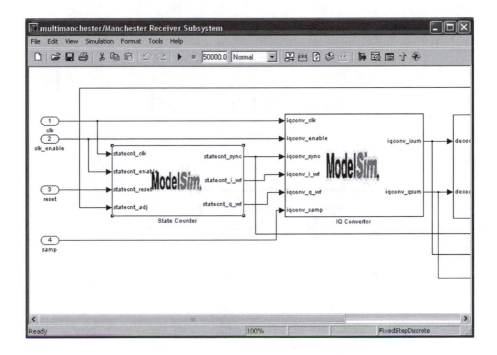

FIGURE I.6
Simulink model that uses the Link for ModelSim VHDL Cosimulation block.

I.1.3.1 Integration of the Hardware Component VHDL Model and the Simulink Model

After the VHDL model is generated and simulated in the ModelSim environment, it is embedded into the Simulink model as a VHDL Cosimulation block. In this integration process a Simulink model window opens, and a model section that should be replaced using the VHDL Cosimulation block is eliminated. Then the Link for ModelSim library is found with the Simulink library browser. It contains three block icons (Figure I.7). The icons of the above-described blocks are transferred to the Simulink model environment, and all requisite links between the model blocks are generated.

The VHDL Cosimulation block has at least one input and one output port. The VHDL Sink block does not have output ports and is intended for signal

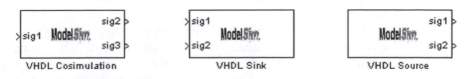

FIGURE I.7
The blocks of the Link for ModelSim library.

FIGURE I.8
The To VCD File Block of the Link for ModelSim library.

analysis of the blocks connected to its input ports. The VHDL Source block does not have input ports and is intended for the generation of the model input signals. In addition to these three blocks, the library has one more block, the VCD File (Figure I.8), whose operation will be described next.

I.1.3.2 Configuration of the VHDL Cosimulation Block

This block is configured by parameter designation in a corresponding dialog window that has the four following bars: the first bar, Ports, describes the input and output ports that correspond to the VHDL project signals (including internal signals; the signals of any hierarchical level can be used), and the simulation time is set for the output signals (VHDL Cosimulation block input ports inherit the input signal simulation time). The second bar, Comm, sets the communication type and its parameters for data exchange between ModelSim and Simulink simulators. The third bar, Clocks, sets the synchronization (from the synchronous pulse front or back edge). The fourth bar, Tcl, sets the Tcl language instructions that a user wants to execute prior to or following the simulation process.

I.1.3.3 Assignment of VHDL Signals to the VHDL Cosimulation Block Ports

To assign VHDL signals in ModelSim, the path names should be given to the VHDL signals that are scheduled to be connected to the VHDL Cosimulation block ports. The signal path name in ModelSim permits the user to visualize the signal hierarchy in the VHDL project. The ModelSim *wave* window is a way of browsing the signal path names (Figure I.9). The VHDL Cosimulation block before and after configuration is shown in Figure I.10.

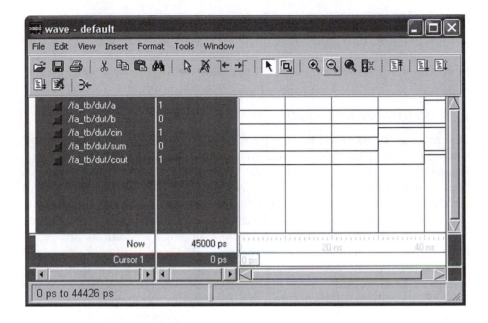

FIGURE I.9
The wave window of the ModelSim system.

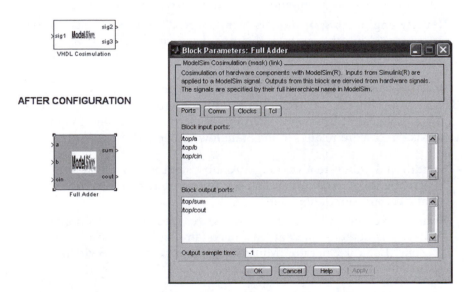

FIGURE I.10
The view of the VHDL Cosimulation block before and after configuration.

I.1.3.4 Start-Up and Testing of a Simulink Cosimulation Model

For start-up and testing, a SimulationStart command should be selected in the Simulink model window; then the model is started up and the detected fault data are preempted.

I.1.4 Reactive Systems–Reactis

One more example of the tendency toward manifestation in the design-for-test area is the product line of Reactive Systems–Reactis: Reactis Simulator, Reactis Tester, and Reactis Validator [11]. Reactis is a set of ESDA tools (embedded-software design automation) that supports the highly efficient development of quality software for various control systems. At present, Reactis has two key components: Reactis Tester and Reactis Simulator.

Reactis Tester automatically generates test sets for the Simulink and State-flow models, providing the software for embedded control systems. These test sets ensure comprehensive coverage of software faults as a function of various test quality metrics and at the same time minimize test redundancy. Each test vector contains both input stimuli for a certain model and the model responses to these stimuli. Such tests can be used for various objectives:

Implementation conformance: To ensure the conformance between the model implementation and behavior, the tests can be applied to the source code of model implementation.

Model testing and debugging: The tests can be applied to the models — to investigate them and correct their behavior if possible.

Reverse engineering of models from source: The model tests can be generated in conformity to the model source codes — to check the conformance between a model and its code.

Reactis Tester employs five fault coverage metrics for the models tested. Basically, the fault coverage metrics record the amount of model syntax forms that are executed at least once. Two initial metrics are related to Simulink, whereas the rest are related to the Stateflow system. These two are:

Conditional subsystem coverage: A subsystem is considered covered if it operates (i.e., the operation condition becomes true) at least once at a certain simulation step.

Branch coverage: A block is considered covered if its conditional behavior is demonstrated at least once (among such blocks are Dead Zone, Logical Operator, MinMax, Multiport Switch, Switch, Relational Operator, and Saturation).

The coverage metrics for Stateflow are determined from the syntax of the system's graphical language. Stateflow diagrams contain states, transition segments, and nodes. Each transition segment, in its turn, has a label, comprising one or more of the following elements: event, condition, condition-induced action, and transition-activated action. These metrics are:

State coverage: This tracks the inputs to the model states.

Condition action coverage: This tracks such transition segments for which the conditional actions related to them were executed at least once.

Transition action coverage: This tracks transition segments for which the actions related to them were executed at least once.

The Reactis Simulator enables users to visualize the model execution as closed and open block diagrams in the Simulink–Stateflow environment. The user interface in the Reactis Simulator is analogous to a conventional debugging interface for programming languages and enables users to conduct simulation in a step-by-step mode, in direct and inverse directions, or using the set of halting points, along with graphic mapping of various fault coverage metrics for various tests. Simulation is conducted either on the Reactis Tester–generated tests or on user tests.

The MathWorks has shown that Model-Based Design using Simulink results in dramatic reductions in development time, cost, and risk, while Xilinx, before there were alternative chip manufacturers, perceived that Simulink was a good platform for FPGA-based DSP design [12]. This trend turned out to be so beneficial that a number of alternative chip manufacturers began using Simulink as their visual design environment. Sometimes other visual design environments are used for similar goals; for instance, National Instruments (NI) makes use of LabVIEW in designing the Xilinx FPGA-based measurement and control platforms [13].

The HDLs, even if they include system-level extensions, cannot efficiently support the rapid modeling and design iteration and algorithmically intensive, large-scale embedded hardware systems. C-based tools also cannot cope with the task on their own. Even large companies do not have a sufficient number of programmers and test engineers to develop and verify manually developed code in C [12].

The SystemC™ and the pure, untimed C++ languages are employed in Mentor Graphics Catapult™ C Synthesis tool, but this is not used at the higher design level, which is MATLAB Domain (Pure Algorithmic Non-Implementation Specific). The next two design levels, lower for the hierarchy, are Untimed C Domain (nonimplementation specific), which uses Standard C language and Timed C Domain (implementation specific), which uses SystemC and Handel-C languages.

Only the lowest design level, RTL Domain (implementation specific), uses the HDLs (VHDL and Verilog) [14].

The Model-Based Design platform is at present the Simulink 6, which has the following properties [12, 15]:

1. Component-based modeling for large-scale systems, including the ability to simulate, test, and implement each design component

independently, as Simulink models are hierarchical and consist of subsystems or components

2. Unified data management for Simulink models and signal parameters across component models, including a graphical model explorer tool
3. Simulink Verification and Validation, which links models to requirements and test cases and identifies untested portions of models
4. The ability to include a MATLAB function and a subset of the MATLAB language in Simulink models and automatically generate embeddable C code
5. A number of new products for Model-Based Design, including Link for ModelSim, for cosimulation and verification of VHDL and Verilog code using Mentor Graphics ModelSim

Within the Model-Based Design framework, specification, design, implementation, and verification are realized via Simulink models. The Model-Based Design key stages are [12] (a) Executable Specifications from Models, (b) Design with Simulation, (c) Implementation with Automatic Code Generation, and (d) Continuous Test and Verification. At the first stage the Simulink models are used as executable specifications for system and component behavior. At the second stage Simulink serves as a platform for multidomain simulation of dynamic systems. Simulink provides an interactive, graphic block diagram environment with a customizable set of block libraries produced by different chip manufacturers for digital signal processing, communications, and control. The fact that Simulink is a MATLAB component provides access to the system tools utilization in algorithm development and data analysis. The hierarchical Simulink models grant all data necessary for the project realization as software or hardware. Simulation in the Simulink environment demonstrates that the project executable specifications are comprehensive and function correctly.

At the third stage, within the framework of a strategic partnership between The MathWorks and chip manufacturers, a code is automatically generated from the Simulink model, eliminating the need for hand-coding and the errors that manual coding can introduce. The generated code is realized in real-time prototyping and deployment in the target system. At the fourth stage, for development-process quality assurance, Simulink models are verified and tested, permitting elimination of certain circuit engineering faults at the early design stage.

Simulink models used as the system models, or *golden references*, serve as the test bench for the hardware or software implementation that is verified and tested during the software or hardware in-the-loop cosimulation.

Work [16] is focused on the functional verification of hardware designs using HDLs (VHDL, Verilog) [17–19] and hardware verification languages (HVLs). In the author's opinion, HDLs cannot efficiently execute functional verification. HVLs have been specially designed for efficient realization of test benches. Among the HVLs are e (Verisity), OpenVera (Synopsis), RAVE

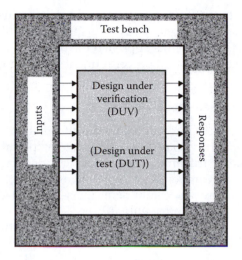

FIGURE I.11
Generic structure of a test bench.

(Forte Design), SystemC Verification Library (Cadence), and Jeda (Juniper Networks).

Application of HVLs permits the realization of a constraint-driven random verification strategy. The notion of verification is not similar to test bench or a series of test benches. It is the process that displays the intent of a design that is preserved in its implementation. Test bench is a simulation code used in generating the deterministic or pseudo-random input sequences to a design and for the observation of corresponding output responses to these input stimuli (Figure I.11).

The difference between verification and testing is as follows. Verification compares the hardware design results as a netlist with design specifications, whereas testing is the verification extension. A manufactured chip is tested for its consistency with the results of an already verified hardware design — a netlist (Figure I.12) [16].

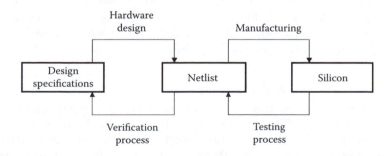

FIGURE I.12
The difference between verification and testing.

This book aims at contributing to the development of test benches for the verification and testing of digital designs, using the Simulink and Stateflow components of MATLAB. In this approach the design of a dynamic-system-representing test bench is based on component-based modeling.

I.1.5 Component-Based Modeling

The system structure can provide valuable information about the system. The majority of modern visual simulation systems permit merely drawing the simulated-system structural diagram using a dedicated graphical processor. The model is composed of the images of blocks comprising the standard set library. Any new component can be assembled from the already existing ones or generated independently, based on the existing mathematical model. Design of new components out of the existing ones generally results in the construction of a hierarchical structural diagram. A modern computer model looks like a definite virtual hardware composed of parallel operating components rather than conventional software (the sequence of operators).

I.1.6 From Classical Dynamic Systems to Hybrid Systems

The term *dynamic system* was first used with the autonomous set of ordinal differential equations, whose right-hand part satisfies the conditions that ensure the solution's existence and uniqueness. Now it is frequently used in cases for which the issue is systems with time-dependable behavior. The hybrid systems, particularly, stand out from the dynamical ones. The hybrid system's behavior consists of the "splicing" of continuous behaviors in the form of piecewise continuous functions. The equations and formulas whose right-hand parts include conditional expressions are sometimes called hybrid equations, emphasizing the fact of coping with the population of various mathematical descriptions whose specific realization is event-dependent.

Hybrid (continuously discrete, event-controllable) systems are the generalization of classical dynamic systems with varying equations, dimensionality, and phase vector composition in different phase–space regions. In the description of hybrid systems, both conventional continuous models (the sets of differential equations) and classical discrete models (finite automata, based on various time models) are applied.

I.1.6.1 Component-Based Models

Hybrid machine operation is reduced to the successive reproduction of classical dynamic models. The continuous constituent of hybrid dynamic-system behavior is defined at each continuity section by the local set of algebraic and differential equations and formulas, whereas the discrete dynamic-system behavior is represented by the machine graph and by algorithmic procedures, executed by the change of states.

A hybrid machine is a graph of transitions, where certain continuous mappings are ascribed to its nodes and certain transition conditions and executed actions are ascribed to its arcs. At present, the Statechart by D. Harel [20,21], which was later canonized in the Unified Modeling Language (UML Reference Manual) standard [22], is a *de facto* standard for formal descriptions of discrete-state machines. The statechart that has certain continuous mappings ascribed to its nodes is called the hybrid statechart. It is a simple and extremely pictorial visual representation of the changes in behavior.

The hybrid machine description "conceals" all possible routes so that each specific route is generated in the solution course as a function of chosen initial conditions and realized events. If the system is a classical dynamic system, it looks like a hybrid machine with one node and lacking transitions.

The difference between the mathematical and the component-based modeling (design of complex engineering systems by means of graphic media — visual simulation systems) is as follows. Mathematical modeling is, first and foremost, the study of global properties of novel theoretical models, using new numerical techniques, and their further realization in the conditions of a chosen mathematical package. The model studied is in this case a unique mathematical object, and its preparation, recording type, and modification are secondary to the investigation of its properties and the search for the novel solution techniques. In designing, in contrast, some novel engineering tasks are solved using the previously investigated models and known numerical techniques. Here, of primary importance are the automation of mathematical model construction, introduction of changes into the existing models, convenience of numerical experiment conduct, and visualization of calculation results. The subjects of investigation are not the global properties, but the specific ones, depending on the constructive parameters of the object being studied.

In design, application of mathematical systems, even in the study of simple engineering systems preset by a single hybrid machine, is quite entangled for two reasons: (1) a hybrid machine is hard to describe when only a textual language form is used, and (2) when solving a current set of equations, the conditions of transition operation should be tracked (the point of a hybrid machine switching from one state into another should be sought). Not all numerical techniques can perform the operation, and very few users can write the required procedures, even if the package language provides such an opportunity.

The important but intricate system can rarely be described by a single classical dynamic system or by a single hybrid machine. Besides, it is very difficult to construct and even more difficult to interpret the aggregate system for a classical dynamic system or a number of aggregate systems for a hybrid machine represented by the required amount of equations and formulas.

This assumption holds true for the structurally intricate simulated objects. The presence of structure is per se indirect evidence of system complexity. A researcher or engineer will find it suitable to focus first on the behavior of individual components, and only after that on their interactions. All engineering systems have natural structure — frequently hierarchical.

It is not improbable that the components of a simulated engineering system are typical for the given applied branch of engineering and that their models (mathematical or structural) have already been developed and listed in the dedicated libraries of standard components. A contemporary approach to design automation is that an engineer just assembles a system developed out of standard components on a virtual test bench in the same way he or she can make it on a real test bench in metal form, without thinking about mathematical models, equations, or statecharts, but assuming that his or her computer model components would interact similarly to their real-world counterparts.

A particular case is systems with dynamic structures, whose component compositions vary with time. Such systems are practically irreproducible in mathematical simulation systems.

Thus, there exists such a modeling domain that should be named component-based modeling, or system designing. In this case the modeling package must first obtain the aggregate system automatically out of the component models and descriptions of their connections, then reduce it to the numerical-solution-suitable form, and only after that, find the system phase paths. This is the difference between the mathematical and the component-based modeling systems.

I.2   Structure of the Book

I.2.1   Models of the VLSI Circuit Fragments

We differentiate the behavioral and the structural models of VLSI (very-large-scale integration) circuits. A behavioral model is a black box that establishes logical connections between inputs and outputs in conformity with the VLSI circuits operation algorithm. At each hierarchical level, a structural model describes the integrity of corresponding elements (blocks) and in-between links. Every basic element of the structural model is described via the behavioral model.

The first section of the book describes the Simulink and Stateflow blocks used (or potentially used) in the model building of VLSI circuit fragments and the test benches used for their verification and testing (Chapters 1 and 2).

I.2.2   Fault Modeling and Simulation

An advantageous solution to the problem of VLSI circuit testing and verification at all design and manufacturing stages ultimately determines crucial characteristics such as fault-free design, performance reliability and stability, low cost, and so on.

Two types of VLSI circuit testing can be differentiated: (1) functional testing executed at all logical circuit development stages, and (2) functional testing of the VLSI circuit samples following their manufacture.

The objective of functional testing is to check the circuit performance validity and its compliance with VLSI circuit specifications. The functional testing specificities are: (1) circuit development is a heuristic process and results in unpredictability of possible faults and their manifestations, and (2) the VLSI circuit is a complex digital machine with an enormous number of internal states that cannot be sorted out during testing. For successful functional testing a hierarchy of functional tests should be created in correlation with the hierarchy of the VLSI circuit–constituting fragments.

In block functional testing particular attention is given to checking the correct links between the in-block functional components and the block intrinsic internal states. Similarly, in functional testing development for VLSI circuits in general, particular attention is paid to checking the correct links between the VLSI circuit blocks. Functional testing is aimed at establishing the manufactured sample compliance with the VLSI circuit project specifications.

Testing is the comparison of behavior of the VLSI circuit and the functional specification (functional testing) or the VLSI circuit gate structure and the structure determined by the project description (gate testing). In functional testing the VLSI circuit is tested for correct operation, but when and how correct functioning should be determined remains unknown. For instance, how many times should a certain arithmetic operation be reiterated for the complete check of a relevant operation-executing circuit? For example, a complete trial of such an n-digit circuit could demand 2^{2n} test vectors (trials).

It is less cumbersome to conduct structural identification of the designed VLSI than complete functional testing. Testing at the gate-structure level uses the gate-structure fault model. Moreover, gate testing and corresponding verification need far fewer test vectors. Reduction of test vectors during structural verification is achieved by finding the signal-passing critical paths for a definite fault.

The VLSI fault is the state induced by the physical defects (opens, shorts, metal bridges, process errors, etc.) of one or more of its elements. The majority of faults results in logical type failures, when a logical zero or logical unity state is set in a circuit and cannot be changed, which violates the regular operation of VLSI circuits. Therefore, the functional control is generally conducted to check the absence of logical-type faults (single constant faults) in VLSI circuits.

The functional testing technique is as follows. Input stimuli are fed to the inputs of a tested VLSI circuit. At given moments of time the states of the VLSI circuit outputs are determined and compared to the reference states for a given elementary check. The comparison results permit the tester to draw conclusions on the VLSI circuit's serviceability with a given input set. The VLSI circuit is tested consecutively at all elementary sets (tests). When a discrepancy between the actual and the expected states of the VLSI circuit outputs is detected, the process is stopped and the VLSI circuit is labeled as faulty. Otherwise, it is thought to be a fault-free circuit.

It is unlikely that most VLSI circuits will obtain a 100% guarantee of fault-free operation. Thus, the construction of acceptable tests is expedient; if the testing time is reduced 100-fold as a consequence of test-time reduction, and at the same time 95% of faults are revealed, it is quite acceptable, since the costs of testing for the remaining 5% can be higher than the costs of a faulty circuit detection and replacement during operation — hence, the importance of quality assessment for the tests devised for VLSI circuit testing.

To estimate test quality, fault simulation techniques are employed (comparison of a fault-free circuit's modeling results and many faulty circuits obtained by alternate introduction of preset faults into a fault-free circuit). The fault is thought to be covered if at least one such test set is available in all the test sets that can detect the mismatch in logical states of a fault-free and a faulty circuit's output [23–25].

Fault modeling and fault simulation for combinational and sequential circuits using Simulink and Stateflow are discussed in Chapter 3.

I.2.3   Testability Analysis Methods

Unlike digital devices, VLSI circuits are nonreparable. The conventional development scenarios are prototyping and sample finishing by means of component replacement or by an alteration in the connection scheme. The basic feature of VLSI computer-aided design is the demand for fault-free design.

VLSI circuits have a limited number of external (primary) outputs, whereas the functions they execute are very complex. As the VLSI circuits' integration level increases, the input to crystal component ratio decreases and the testing task becomes more intricate, as we feel the shortage of external outputs for signal feeding and observation at various circuit points. The number of observation points is only limited by the VLSI circuit's input and output terminals. Hence, significant attention is paid to the development of novel techniques of test quality assurance, test generation, and testable design.

As the VLSI circuit's integration level increases and the gate dimensions decrease, the challenge of testing the finished VLSI circuits is faced, since the contact control sites should be eliminated because of the deficient crystal area. As a result, numerous logical blocks are beyond the reach of testing, and as a consequence, testing of VLSI circuits becomes more sophisticated and expensive. Testing of finished VLSI circuits is the verification of a crystal-based project.

These problems raise the issue of designing testable (controllable) VLSI circuits. Testability is a quantitative measure of the ease of testing. Two essential notions are used in the analysis of VLSI circuit testability: observability of a circuit node and controllability of a circuit node. The controllable node is connected to the VLSI circuit's external output, whose logical state can be easily defined. The possibility of a controllable-node forced setting to the predetermined logical state is ensured.

VLSI circuits with high testability have the following distinctive features:

1. The circuit is initialized via the major input buses (known signal values are set at all circuit nodes).

2. A short sequence of input signals can be used to control the VLSI circuit's internal state.

3. There is no need to use all major inputs for the correlation of internal signals.

4. Internal states and signal internal values are defined unambiguously from the data at major outputs or in special tests points.

The first three features reflect the controllability level, while the fourth reflects the VLSI circuit's observability level. In other words, to make the circuit completely testable, its internal states should be easily controlled and observed. Controllability and observability of all VLSI circuit signals reflect the testability level of VLSI circuits as a whole.

Programs for VLSI circuit testability analysis at the design stages comprise the verification facilities, though during the VLSI design the programs per se are not appreciably requisite. However, in combination with alternative design techniques, the testability analysis assists the designer in elucidating the VLSI circuits' problematic points [23,24].

The models for VLSI testability analysis constructed with the use of Simulink and Stateflow are discussed in Chapter 4.

I.2.4 Automatic Test-Pattern Generation Process

The essence of the test synthesis (generation) problem is the development of test sets that assist in detecting all or most faults in VLSI. Several test generation techniques have won recognition:

1. Rot's algorithm (D-algorithm)
2. The PODEM algorithm
3. The probabilistic technique (fault modeling time is cubically dependent on the number of circuit elements. Therefore, the techniques that make allowance for the probabilistic factors are being extensively developed) [23–25]

The models of such deterministic algorithms constructed with the use of Simulink and Stateflow are discussed in Chapter 5.

I.2.5 Timing Verification

Verification facilities provide the testing of design procedure execution, imitation simulation, and realization of formalized testing procedures for VLSI circuit projects (from function descriptions up to topology). The verification is based on VLSI circuit simulation at any representation level.

Functional simulation is aimed at clarifying the VLSI circuit's serviceability at the block level. In this process the behavioral model is a finite state machine and can contain data on the modeled block time parameters.

Gate-level simulation can be done with (asynchronous simulation) or without (synchronous simulation) taking into account the delays. Gate-level verification provides the designer with data on logical function execution in the project (synchronous simulation), on the project time parameters (asynchronous simulation), and on critical signal passing paths. Timing verification is based on two timing analysis procedures in digital VLSI circuits:

1. Logical circuit analysis as a whole: analysis of all signal passing paths. The longest (critical) path is memorized; it helps in assessing the VLSI circuit's operation within the entire environmental parameter scope.
2. Analysis only of critical signal passing paths that give the maximal delay. Critical paths are determined from the event tables.

At early design stages, the timing characteristics of VLSI circuits cannot be correctly evaluated. Such information can be obtained only after the topological design is completed. In many cases the propagation delay depends on the number of gates connected to the outputs of similar gates. During verification, this peculiarity is accounted for in the input and output fanout factor. For better accuracy, timing analysis is made for minimal and maximal delays, thus providing the data on delay propagations at the logical block outputs. Infrequently, during timing verification, the necessity arises of establishing whether the output signal is present when a signal is fed to each input and whether the investigated section is inverted. For complex projects, such analysis is information dependent.

Timing verification assists in finding and correcting the potential errors that affect the project's dynamic parameters. Timing verification based on the analysis of critical-signal passing paths is most beneficial for asynchronous circuits.

A designer can specify a critical section and calculate the corresponding delay, or he or she can check all sections and find precisely that one whose delay exceeds the specification requirements. Such verification can be made for any input signal. Having detected the critical path with delayed signal propagation exceeding the specification requirements, the designer can redesign it. The correct approach demands examining all signal-passing critical paths where the delays are higher or lower than the specifications.

The verification based on critical-path analysis is generally used in two cases: (1) before topological design, when crude estimates of the circuit load's nature are available, and (2) following topological design, when the factual load pattern can be extrapolated and more accurate delay and alternative timing parameter values can be determined.

In practice, it is insufficient to show that the VLSI circuit performs the function assigned by specifications and that its topology corresponds to the logical circuit. It is also necessary that the VLSI project have a safety factor (i.e., a safety margin for the scattering of signal propagation delays in the circuit induced by differing lengths and specific resistances of topological circuits, for the technological parameters scattering and for the environmental factors influence). Such analysis is called verification and validation (V & V) and is made by computer simulation of major tests. It results in the development of logical simulation techniques that map the circuit processes with increased accuracy. Logical simulation with regard for the topological circuit delays permits the designer to assess the project's immunity to technological parameter scattering (specific resistance of topological layers, threshold transistor operation voltage), to power source voltage, temperature, and so on.

The influence of signal propagation delays in topological circuits on logical circuit serviceability is evaluated as follows:

1. Each circuit is represented as an resistor capacitor (RC)-tree with one (or more) source and many receivers.
2. The delay in signal propagation from each source to all circuit receivers is calculated by assessing the delay upper and lower borders and their subsequent averaging.
3. The delays obtained are taken into account during logical simulation [26].

Timing verification using Simulink models is discussed in Chapter 6.

I.2.6 System and Embedded Core Testing

Despite the process's apparent simplicity, functional testing has one extremely grave contradiction — between test quality and VLSI circuit testing time, which ultimately determines the test costs. Therefore, as the VLSI circuits become more and more complicated, the necessity of acceptable testing techniques in development and generation is felt.

Possible ways to increase the efficiency of VLSI circuit functional testing are:

1. Development of design techniques for testable VLSI circuits aimed at the enhanced testability level in VLSI circuits
2. Development of embedded self-testing techniques in VLSI circuits
3. Development of computer-aided synthesis and test minimization techniques
4. Development of test quality analysis techniques (i.e., assessment of testing completeness in VLSI circuits using some given tests)

The methods of testable VLSI circuit design relate to the circuits that allow no probing and serviceability restoration. Therefore, their operation capacity can be assessed only by comparison of the totality of logical zeros and logical unities at external outputs obtained in response to an input test.

To increase the observability and controllability of the circuit's internal nodes, various design techniques for testable VLSI circuits are applied.

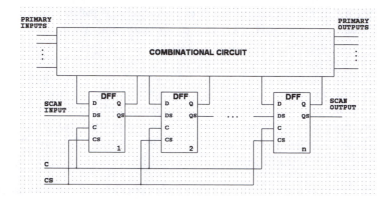

FIGURE I.13
The diagram of flip-flop unification into the shift register for scanning path organization.

The most widely used is the scanning path technique and its modifications, by which regular memory elements of VLSI circuits can be either reconfigured into the shift register or replaced by the shift register (Figure I.13). As a result, the process of data input and output to VLSI circuits becomes simpler. In doing this, each circuit flip-flop can operate in two modes: (1) as a regular flip-flop, and (2) as a flip-flop operating in shift register.

Generally, three testing modes are feasible:

1. Testing of the scanning path itself (the shift register can be checked separately, by loading arbitrary sequences of logical zeros and logical unities via the scanning input and subsequent reading of the same sequence via the scanning output)

2. Assignment of the circuit initial state based on the scanning path and its subsequent testing via external inputs

3. Reading of the combinational circuit state

One variety of scanning path technique is the level sensitive scan design (LSSD) method intended for the design of VLSI circuit crystals. This uses the dedicated memory elements, where the states are changed by the synchronization signal level rather than by its front. The locking flip-flop diagram (Figure I.14) includes two flip-flops (D1, D2). D1 is a common memory element. In regular operation conditions, controlled by the system synchronous signal C, the shift control signals, CS1 and CS2, are equal to zero and DS input and D2 flip-flop are not used. With absent synchronization, when a logical unity is set at CS1 input, the data from D1 are rerecorded in D2 and preserved there at CS2 = 0. Control signals, CS1 and CS2, are not obscured and are fed to all memory elements via the VLSI special terminals. From QS output, the data are fed to the DS input of the following memory element, generating the scanning path. The LSSD technique is particularly

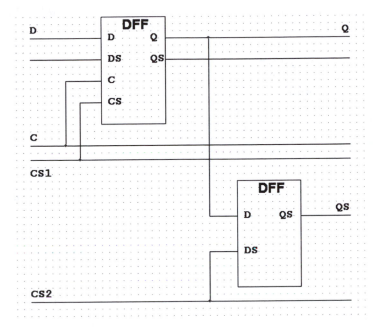

FIGURE I.14
Logical memory scheme for scanning path organization by the LSSD technique.

suitable for metal oxide semiconductor (MOS) circuits, since locking flip-flop D2 can be implemented on one transistor only.

If these testing techniques are employed, all flip-flop inputs are accessible, and all outputs are observable. Compilation of tests is essentially simplified. Testing using the scanning techniques permits the designer to assess the VLSI circuits' serviceability, to reduce their costs, and to upgrade testing quality.

The scanning techniques have the following drawbacks:

1. Deterioration of VLSI dynamic characteristics caused by scanning circuits
2. Doubling of flip-flops
3. Demand for appreciable crystal areas destined for the layout of scanning control circuits
4. Three or four VLSI external outputs

Thus, in practice, the incomplete shift register is used that pools up to 30% of the most complex testable flip-flops.

Wide acceptance has been gained by self-testing VLSI circuits and embedded-testing VLSI circuits. These additional testing circuits operate during VLSI circuit performance or in the intervals between useful information processing (direct mode and scanning mode). The self-testing VLSI circuits and embedded-testing VLSI circuits are designed within the framework of general designing, using the verification facilities.

The existing test synthesis systems are sufficiently efficient for combinational circuits but display poor performance at arbitrary sequential circuits (here the fault nature depends not only on the input signals but on the circuit's internal state as well). The cardinal way out is the transition to design techniques for testable VLSI (LSSD, built-in self-test, etc.) [23–25,27]. In doing this, the test development order involves the following steps:

1. Debugging of the VLSI circuit's logical circuit
2. Assessment of the circuit testability by seeking the controllability (observability) tree
3. Prediction of test quality, testing costs, and fault missing costs
4. With an unsatisfactory circuit testability estimate, isolation of flip-flops to be potentially included in the shift register, introduction of changes to the VLSI circuit, and transfer to step 1
5. With a satisfactory circuit testability estimate, automatic test generation based on the circuit-testability analysis results

Application of Simulink models for the achievement of certain results in this area is discussed in Chapter 7.

References

1. Morris K. Destination DSP. Methodologies for signal processing success. *FPGA and Programmable Logic Journal* (www.fpgajournal.com/articles/20041130_dsp.htm).
2. Mascarin A. MATLAB & Simulink Accelerating Implementation of DSP designs on FPGAs (www.xilinx.com/events/docs/esc_sf2001_tmwpres.pdf).
3. Lam S. Implementing DSP algorithms in FPGAs. *Xcell Journal*, 51, 56–59, 2004.
4. System Generator for DSP (www.xilinx.com/ise/optional_prod/system_generator. htm).
5. Rabaey J.M., Chandrakasan A., Nicolic B. *Digital Integrated Circuits: A Design Perspective* (2nd ed.). Pearson Education International, Upper Saddle River, NJ, 2003.
6. Dhanani S., Zack S. Enabling low-cost DSP co-processing with Spartan-3 FPGAs. *Xcell Journal*, 49, 88–90, 2004.
7. Denning D. Accelerate and verify algorithms with the XtremeDSP Development Kit-II. *Xcell Journal*, 49, 82–84, 2004.
8. Cigan E., Lall N. Integrating MATLAB algorithms into FPGA designs. *Xcell Journal*, 53, 73–75, 53.
9. *Altera DSP Builder User Guide*. Altera Corp., San Jose, CA, 2003.

10. *Link for ModelSim for Use with MATLAB and Simulink User's Guide*. The MathWorks, Inc., Natick, MA, 2004.
11. *Reactis™ User's Guide V2003b5* (V2004). Reactive Systems, Inc., Falls Church, VA, 2003.
12. Karnofsky K. Simulink brings model-based design to embedded signal processing. *Xcell Journal*, 51, 66–69, 2004.
13. Trimborn M. Building high-performance measurement and control systems with FPGAs. *Xcell Journal*, 53, 90–92, 2005.
14. McCloud S. Algorithmic C synthesis optimizes ESL design flows. *Xcell Journal*, 51, 46–51, 2004.
15. Using Simulink: Dynamic System Simulation for MATLAB. The MathWorks, Inc., Natick, MA, 2004.
16. Bergeron J. *Writing Testbenches: Functional Verification of HDL Models*. Kluwer Academic Publishers, Dordrecht, The Netherlands, 2003.
17. Ashenden P.J. *The Designer's Guide to VHDL* (2nd ed.). Morgan Kaufmann Publishers, San Mateo, CA, 2002.
18. Ashenden P.J., Peterson G.D., Teegarden D.A. *The System Designer's Guide to VHDL: AMS (Analog, Mixed-Signal, and Mixed-Technology Modeling)*. Morgan Kaufmann Publishers, San Mateo, CA, 2003.
19. Palnitkar S. *Verilog HDL: A Guide to Digital Design and Synthesis*. Prentice Hall PTR, New York, 1996.
20. Harel D. Statecharts: a visual formalism for complex systems. *Science of Computer Programming*, 8(3), 231–274, 1987.
21. Harel D. On visual formalisms. *Communications of the ACM*, 31(5), 514–530, 1988.
22. Rumbaugh J., Jacobson I., Booch G. *The Unified Modeling Language Reference Manual*. Addison-Wesley-Longman, Reading, MA, 2001.
23. Bushnell M.L., Agrawal V.D. *Essentials of Electronic Testing for Digital, Memory, & Mixed-Signal VLSI Circuits*, Kluwer Academic Publishers, Dordrecht, The Netherlands, 2004.
24. Smith M.J.S. *Application-Specific Integrated Circuits*. Addison-Wesley, Reading, MA, 1997.
25. Crouch A.L. *Design-for-Test for Digital IC's and Embedded Core Systems*. Prentice Hall PTR, New York, 1999.
26. Nekoogar F. *Timing Verification of Application-Specific Integrated Circuits*. Prentice Hall PTR, New York, 1999.
27. Parker K.P. *The Boundary-Scan Handbook* (3rd ed.). Kluwer Academic Publishers, Dordrecht, The Netherlands, 2003.

About the Author

Dr. Evgeni Perelroyzen earned his M.Sc. in 1976 from the Polytechnic Institute of Penza (USSR). The same year he started working in the industries dealing with design-for-test. He earned his D.Sc. in 1982 from the Riga Polytechnic Institute (Latvia). Dr. Perelroyzen has been an assistant professor and an associate professor. Since 1996 he has lived in Herzeliya, Israel.

He is a senior teaching fellow at the Department of Mathematics (University of Haifa) and the Faculty of Industry Engineering and Management (Technion) and an adjunct lecturer at the Department of Computer Science and Engineering (Hebrew University in Jerusalem). He is the CEO of Perelroyzen EDA Planning, Ltd. He is the author of more than 70 scientific publications and books.

Dr. Perelroyzen can be contacted at evgeni-p@smile.net.il.

1

Simulink®: Dynamic System Simulation for MATLAB®

1.1 Introduction

Simulink is treated as the totality of methods and facilities for the automation of modern systems development. The most essential Simulink applications are rapid prototyping and rapid applications development.

Simulink is the extension of MATLAB and assists in reducing the design period, in quality upgrading for physical model development, and process simulation in such systems. In Simulink the nature of software demands has been changed drastically, and the model graphic description as the system block diagram has been developed for computation control. Representation in the form of a block diagram, in most cases, does not need code writing, owing to the blockset library used for the solution of individual applied tasks.

Simulink represents a graphic form of a simulation language, with classes represented by blocks. The classes of destinations are grouped into extensive libraries that can be used for the generation of new system models. Simulink permits the designer to unite block diagrams into complex blocks, ensuring the hierarchical representation of a model structure [1,2].

1.1.1 Hierarchical Systems

Using blocks as the analog of any real component is at the heart of all simulation systems. Blocks can be coupled, forming plane-located functional diagrams. Any functional diagram can be treated as a complex block. Such complex blocks with their internal structures can be coupled again, constructing hierarchical, multilevel systems. A block is an independent, internally functional element that interacts with the ambient world (with other blocks) via a preassigned set of interface variables. A block is generally depicted as a rectangle with corresponding interface variables (inputs, outputs, contacts).

The world of real physical objects possesses the set of standard components and numerous devices using them as ready-made elements. In simulation packages they are matched by the class libraries and by block functional diagrams of the designed devices, constructed from the existing class specimens. Each functional diagram allows differentiation of the block types (classes) used for its construction and specific blocks (class specimens). Functional diagrams usually show the interconnected elements. The connections convey information on equivalent interface variables and the meaning of this equivalence.

A graph is a mathematical model of a system composed of interconnected elements. A graph can be disoriented (only the contact blocks are used) or oriented (the input–output blocks are used), depending on oriented or disoriented connections. When both contact and input–output blocks can be used on functional diagrams, the graph has two types of arcs: oriented and disoriented.

Representing the system as interconnected, simple, and lacking intrinsic internal structural elements, we can proceed further and examine each such block as a new subsystem, if necessary. We can construct a set of subsystems, a new enlarged element base, by uniting all blocks into the nonintersecting groups (subsystems). Proceeding with the process, we can construct a new element base and repeat it until only one subsystem remains in the set of subsystems. It is this system that will be hierarchical. If N steps were repeated, we would have an N-level hierarchical system described by the appropriate hierarchical tree. Instead of using the hierarchical tree, each hierarchical level can be depicted on a separate leaf, where each subsystem is represented by the rectangle with a subsystem name. These enlarged elements assist in constructing the multilevel functional diagrams.

1.1.1.1   Blocks and Connections: Various Approaches

Simulink uses preassigned blocks, which are graphic analogs of electronic circuits. A user can assemble the elementary applied block out of standard universal blocks or must write a new software code in a procedural language. The critics of this approach call the relict of analog computer simulation. The approach's advantages are:

1. It is user friendly for an unqualified user.
2. There is no need for a complex translator.
3. It permits to effective interpretation of the models.

External variables are clearly subdivided into input variables, which can only be changed beyond the block, and output variables, which can be changed only inside the block. Such an approach is frequently called the block approach, and it demands explicit description of casual relations and use of oriented blocks.

1.1.1.2   Oriented Blocks and Connections

Use of oriented blocks assumes that composing the behavior-describing equations causes constraints to be imposed on the employment of phase-vector variables.

The state variable is such an internal block variable that can obtain its new value only inside the block. The state variable can be located on the left-hand side of the assignment operator in transition actions and on the left-hand side of formulas and differential equations; it can also be a useful variable in algebraic equations.

Input is an external variable whose value can be assigned only from the outside. Output is an external variable whose value can be assigned only from the block inside. Functional relations that connect the inputs and outputs of oriented blocks are oriented by virtue of properties of these external variables.

1.1.1.3   *Description of Hybrid Systems*

Simulink uses various techniques for the description of different hybrid behavior types. In the simplest case, when the set of equations is constant, and the values of certain variables are changed jumpwise, it is sufficient to include some discrete blocks operating in combination with the continuous ones into the functional diagram. The most common case of a discrete block is the Harel's Statechart, assigned in Stateflow® [3]. When the equation's composition and number are changing, a functional diagram is assembled that corresponds to the joint of all equations, and special blocks are used that operate like switches for connecting the needed and disconnecting the unneeded circuit sections at the appropriate moments. The switches are controlled by the discrete variables whose values can be varied specifically by the statecharts.

Following drag-and-drop operation, the class graphic image is imported from a library of classes to a container (model window), and a class specimen with default parameters that can be changed later, is generated. A user has access to the class specimen (block) input, output, and parameters. All other details are hidden (encapsulated) inside. A new user class can be assembled only out of the existing classes and designed as a subsystem library class.

The presence of continuous and discrete blocks in Simulink and their potential concurrent application shows that hybrid systems are in general the Simulink simulation objects. Stateflow (Chapter 2) is used for the description of systems with jumpwise changes of parameters. Various switches such as the Switch Block are used in Simulink for the description of systems with differing behaviors. Also, Simulink provides some dedicated blocks that respond to special events and produce special signals (Enable, Trigger, Hit Crossing).

Thus, Simulink is an interactive environment for simulation and analysis of a wide scope of dynamic systems (continuous, discrete, and hybrid [linear and nonlinear]) with the use of block diagrams as graphic means [1,2].

1.2   Creating a Model

Model description in Simulink consists of a block diagram that includes blocks, which in turn are the mathematical models of elements from the continuous, discrete, and discrete–continuous (hybrid) systems.

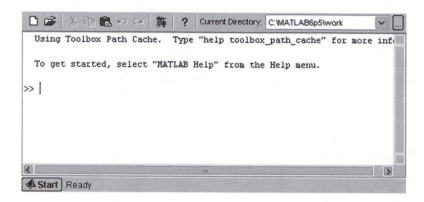

FIGURE 1.1
Simulink icon on the MATLAB Tool Bar.

A block diagram describes a mathematical model whose functional subsystems and their interconnections are fully defined. The models of system elements, in combination with their interactions, permit the designer to construct the dynamic system's mathematical model. During system modeling, each block model inside the block diagram (system model) is used in corresponding calculations again and again. Application of Stateflow permits the addition of procedural (behavioral) blocks to Simulink models, but this additional facility is not based on the block diagram concepts.

Let us discuss model construction as a block diagram. Simulink is initiated from MATLAB by putting the mouse pointer on the Simulink icon on the MATLAB toolbar (Figure 1.1). This opens the Simulink Library Browser window (Figure 1.2). The system new window can be opened by clicking the left mouse button on the Create a New Model icon in the current window toolbar. The Model instruction in the NewFile menu or the combination Ctrl+N keys can also be used. The default window name is Untitled (Figure 1.3). It is in this window that the system model will be constructed as a block diagram.

The toolbar icons permit the use of basic instructions for system model generation and starting simulation (described in detail in [1,2]).

The state panel at the lower part of the window displays a scaling factor of the system block diagram and current simulation time (displayed during simulation). Double-clicking the mouse left button on the icons found in the left side of the Simulink Library Browser window, or by single-clicking on the plus sign makes the list of blockset libraries appear. Clicking the left mouse button on the row with the blockset library name displays a list of blocks in a given library in the right side of the window (Figure 1.4).

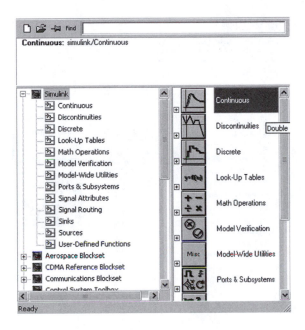

FIGURE 1.2
Simulink Library Browser window.

In Simulink the object models are grouped according to their functional feature:

1. Signal sources (grouped in the blockset library)
2. System element models
3. Imaging devices for the simulation results (located in the blockset library)

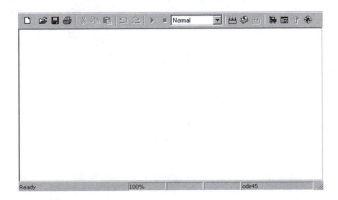

FIGURE 1.3
Empty (new) model window.

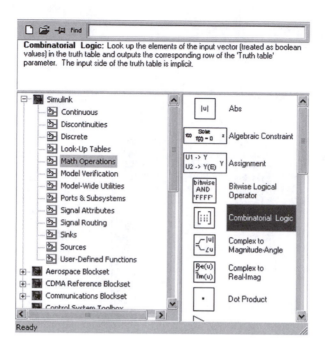

FIGURE 1.4
Simulink Block Library window.

The source output signals act on the system model as its block diagram. The simulation results are recorded by imaging devices as graphs or as the combination of numbers at the display and can be saved in special files. To construct the model block diagram in the default name window, untitled, block images are exported from the blockset library to the model window (Figure 1.5).

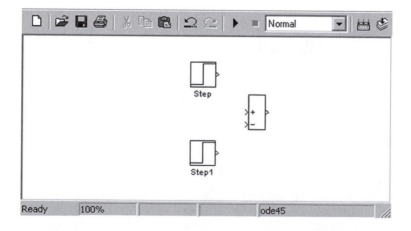

FIGURE 1.5
Dragging a block from a Simulink Block Library.

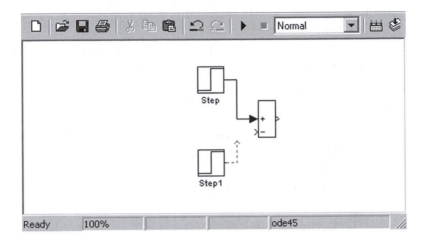

FIGURE 1.6
Drawing a signal line.

Then the connection lines are made by pointing the cursor on the block output port, which makes it look like a cross. After that, with the mouse held down, the cursor is moved to the block input port, taking the shape of a double cross. The connection line ends in an arrow indicating the signal transmission direction (Figure 1.6).

Block images can be edited by changing dimensions, rotating, copying, performing name operations, and so on. To change the block image size, the mouse left button should be clicked on the block image. This results in a rectangular box with markers at its corners. Then the mouse pointer is placed on the selected corner, the left mouse button is clicked and held, and the selected corner is moved to conform to the block image's new size. As the block's requisite size is achieved, the mouse left button is released. Markers are removed by a click of the left mouse button outside the block image.

To rotate blocks, the necessary block is chosen and the Rotate instruction from the Format menu in the model window is used.

To copy a block image, the mouse pointer is put on the block image and the right mouse button is clicked. With the button clicked, the rectangular box with markers is exported to the necessary position and the button is released; to remove the markers, the left mouse button is clicked outside the block image. An alternative copying technique is the use of the Copy Edit (Ctrl+C) Paste Edit (Ctrl+V) instruction.

Listed below are the basic operations with objects and brief descriptions using the following definitions:

Drag: The left mouse button is clicked and held, the object is selected, and the object image location can be assigned by the mouse cursor moves.

Shift-Click: The Shift key is clicked and held while simultaneously clicking the left mouse button.

Shift-Drag: The Shift key is clicked and held, and the object is moved, with the left mouse button being clicked and held.

Ctrl-Drag: The Ctrl key is clicked and held, and the object is moved, with the left mouse button being clicked and held.

1. **Select object (block or signal line):** The object image is selected (the mouse cursor is placed on the object image and the left mouse left button is clicked).

2. **Select another object:** The Shift-Click operation is performed for the added object.

3. **Select with bounding box:** The left mouse button is clicked in a corner of a selecting rectangle, the mouse button is held, and the mouse is moved to encompass all selected objects inside the rectangle.

4. **Copy block from library or another model:** A block image is selected and exported to the model window.

5. **Flip block:** A block image is selected and the Flip Block instruction from the Format menu of the model window (or the combination of Ctrl+I keys) is used.

6. **Rotate block:** A block image is selected and the Rotate Block instruction from the Format menu of the model window (or the combination of Ctrl+R keys) is used.

7. **Resize block:** A block image is selected, and the mouse cursor placed onto the outline frame corner is moved to assign the necessary image sizes.

8. **Add drop shadow:** A block image is selected, and the Show Drop Shadow instruction out of the Format menu of the model window is used.

9. **Edit block name:** The text box of the block name is selected, and the block name is edited in the outline frame.

10. **Hide block name:** A block image is selected, and the Hide name instruction out of the Format menu of the model window is used.

11. **Delete object:** An object image is selected and the Clear instruction out of the Format menu (or the Delete key) of the model window is used.

12. **Copy object to clipboard:** An object image is selected and the Copy instruction out of the Edit menu (or the combination of Ctrl+Y keys) of the model window is used.

13. **Cut to clipboard:** An object image is selected, and the Cut instruction out of the Edit menu (or the combination of Ctrl+S keys) of the model window is used.

14. **Paste from clipboard:** The Paste instruction out of the Edit menu (or the combination of Ctrl+V keys) of the model window is used.

15. **Draw signal line in segments:** The signal line end is moved to the first break point, the left mouse button is released and pressed, the signal line end is moved to the second point, and so on.

16. **Branch from signal line:** The Ctrl-Drag operation is made at the branching point, or the right mouse button is used instead of the pressed left button and Ctrl key.

17. **Label signal line:** The left mouse button is double-clicked on the signal line, and the text is imported to the outline frame.

18. **Move signal line label:** The Drag operation is applied to the declaration text box, and the declaration fixed state with respect to the signal line segment can be assigned by the mouse cursor move.

19. **Copy signal line label:** The Ctrl-Drag operation is applied to the declaration text box, and the declaration copy fixed state with respect to the signal line segment can be assigned by a mouse cursor move.

20. **Add annotation to model:** The left mouse button is double-clicked to indicate the text position, and the text is imported to the outline frame.

The process of copying a block image is initiated by selecting the Start instruction in the Simulation menu. To view the simulation results, double-click the right mouse button on the Scope Block image to open the block dialog window that contains the chart of the process being studied as a function of simulation time (Figure 1.7).

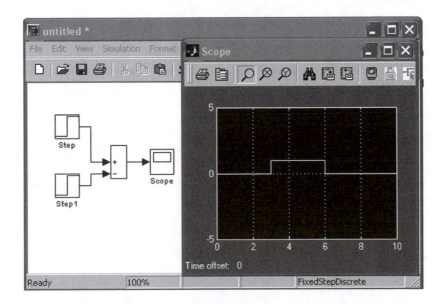

FIGURE 1.7
Scope display after executing the model.

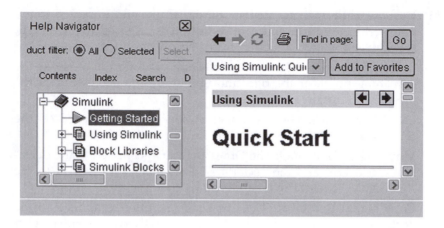

FIGURE 1.8
The Simulink Help window.

Simulink gives the user the opportunity of using the Help subsystem. The help data are displayed by the Simulink Help operation in the Help menu of the model window and by clicking the Help button in the Block Parameters dialog window or in the Simulation Parameters window, called by the same instruction in the Simulation menu (Figure 1.8).

1.3   Running a Simulation

Simulation parameters and options are set either by the Simulation menu instructions or by the toolbar icons of the model window. For simulation startup, the Start instruction is chosen in the Simulation menu or with the Start icon. For simulation interruption at an arbitrary moment, the Stop instruction from the Simulation menu or the toolbar Stop icon is used. Simulation can also be initiated from the MATLAB command line [1,2].

During simulation, certain parameters can be edited (for instance, the changes in the enhancement factor of Gain Blocks or selection of signal lines for the connection of the Floating Scope Block devised for simulation visualization [1,2]).

1.4   Analyzing Simulation Results

The charts of simulation-resulting output signals can be obtained by one of the following three techniques.

1. Use of the Scope Block or XY Graph Block (the finite numerical values of output signals are recorded using the Display Block). The Scope Block screen permits the designer to increase the output data intervals that are of interest to a user or to save them in Workspace. The XY Graph Block permits comparison of two definite signals (see below).

2. Recording of output signals using return variables and MATLAB plotting instructions [1,2]. Certain variables can be specified in the Simulation menu at the Workspace I/O panel of the Simulation Parameters window. Then the corresponding charts can be printed using the MATLAB plot (tout,yout) function.

3. Recording of output signals to the workspace using the ToWorkspace Block and MATLAB plotting instructions [1,2].

1.5 Subsystems: Using Masks to Customize Blocks

The block hierarchical approach to Simulink simulation provides for simulated system partitioning into hierarchical levels (subsystems). At the nest level the simulated system is examined not as a whole but as separate subsystems. The subsystems can be examined as block diagram elements intended for multiuse [1,2]. Simulink provides two methods of subsystem generation:

1. Use of a Create subsystem instruction in the Edit menu of the model window.

2. Use of the Subsystem Block from the Ports&Subsystems Library.

For instance, when the second method is used, the Subsystem Block from the Ports&Subsystems Library should be imported to the new model window and the mouse left button double-clicked on the block image. This will open the subsystem window. Then, the subsystem block diagram is shaped. All subsystem-incoming signal lines should be connected to the output ports of In1 Blocks, whereas all subsystem-outgoing signal lines should be connected to the output ports of Out1 Blocks. If necessary, the names of In1 and Out1 Blocks can be changed to identify each input and output variable. When the shaping of the subsystem structure is completed, its window closes. The subsystem is a part of the model wherein it is created, and it will be saved together with the main model.

Masking is a Simulink procedure that permits the designer to realize the abstraction concept, separating the essential behavioral features and properties from the specific realization traits. Masking allows operating with a subsystem like with a common block. The masked block has its own graphic image and dialog window with parameter settings, similar to the dialog windows of common blocks from the Simulink libraries. The following operations permit the designer to create a masked block [1,2].

1. A subsystem is created.
2. A Subsystem Block is selected, and the Mask Subsystem instruction is chosen from the Edit menu of the model window.
3. With the Mask Editor dialog window, all necessary comments and a masked block dialog window are generated (the block's individual graphic image can be added, if necessary).

1.6   Reference Blocks

Certainly, there is no possibility of examining all Simulink blocks in this book. We demonstrate here a number of blocks that are employed for the design of test benches and VLSI (very-large-scale integration) circuit fragments (design-under-test [DUT]), as well as blocks that can be potentially employed for this goal.

1.6.1   Continuous Library

This library includes the blocks for constructing continuous system models.

1.6.1.1   Transport Delay Block

This block calculates the output variable that is the input variable delayed for a given period (Figure 1.9).

1.6.1.2   Variable Transport Delay Block

This block calculates the output variable that is the input variable fed to the upper input port and delayed for the period set by the signal fed to the lower input port (Figure 1.10).

1.6.2   Discrete Library

The Discrete Library includes the blocks for constructing discrete system models.

1.6.2.1   Discrete Filter Block

This block shapes the model as a discrete system transfer function, represented by the fraction rational function (relation of two polynomials).

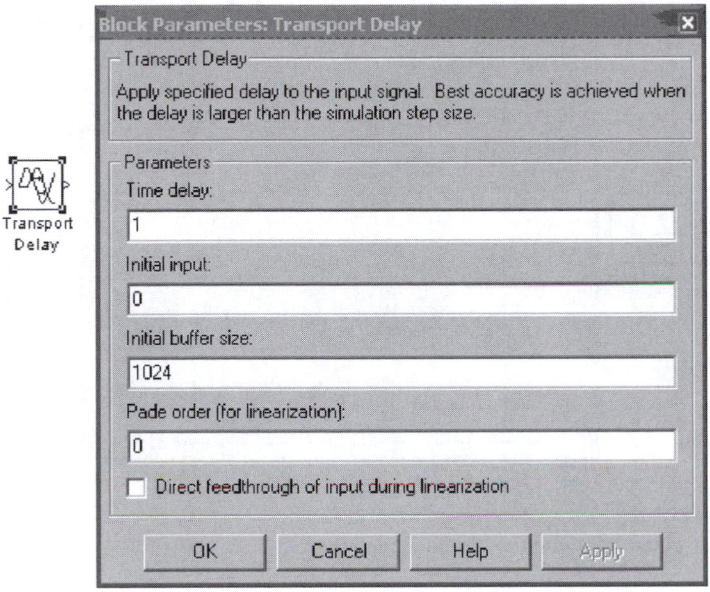

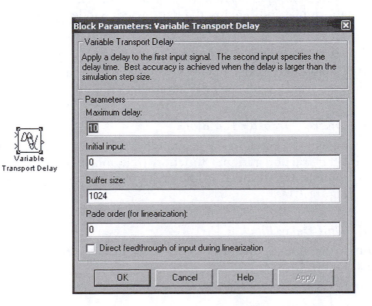

FIGURE 1.9
The Transport Delay Block and the Block dialog box.

FIGURE 1.10
The Variable Transport Delay Block and the Block dialog box.

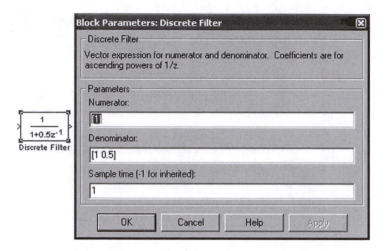

FIGURE 1.11
The Discrete Filter Block and the Block dialog box.

The polynomials are arranged in the order of increasing powers of z^{-1} variable (Figure 1.11).

1.6.2.2   Discrete-Time Integrator Block

This block calculates the meaning of a certain integral from the numerical integration formulas indicated in the Parameters window (Figure 1.12).

1.6.2.3   Discrete State-Space Block

This block shapes the model as the state equations for a multidimensional linear discrete system with constant parameters (Figure 1.13).

1.6.2.4   Discrete Transfer Fcn Block

This block shapes a model as a discrete system transfer function represented by the fraction rational function (relation of two polynomials). The polynomials are arranged in the order of increasing powers of the z variable (Figure 1.14).

1.6.2.5   Unit Delay Block

The output signal takes the input signal value at the previous quantization moment (Figure 1.15).

1.6.3   Discontinuities Library

This library includes the blocks whose outputs are not continuous functions of the blocks' inputs.

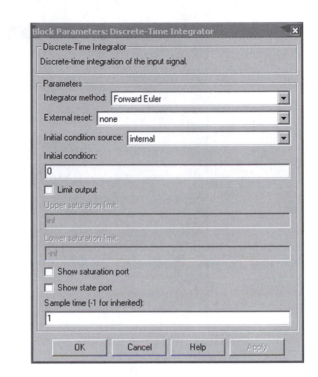

FIGURE 1.12
The Discrete-Time Integrator Block and the Block dialog box.

1.6.3.1 Hit Crossing Block

This block permits the designer to determine the moment of certain level crossing by the signal (Figure 1.16).

1.6.4 Link for ModelSim Library

This library permits creation of the Simulink–ModelSim link (see Introduction).

1.6.4.1 The VHDL Cosimulation, VHDL Sink, VHDL Source, and To VCD File Blocks

These functions serve as the integration of hardware component models in ModelSim and alternative Simulink-located general model constituents (see details in Introduction) (Figure 1.17) [4].

1.6.5 Look-Up Tables

The library includes the interpolation performance blocks.

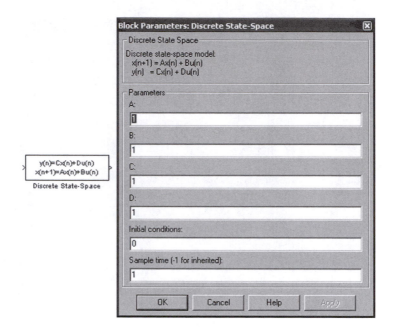

FIGURE 1.13
The Discrete State-Space Block and the Block dialog box.

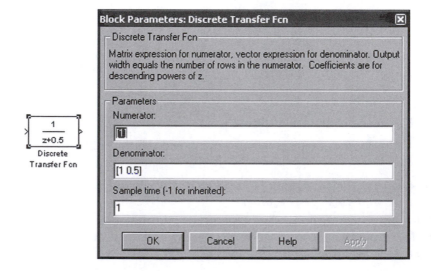

FIGURE 1.14
The Discrete Transfer Fcn Block and the Block dialog box.

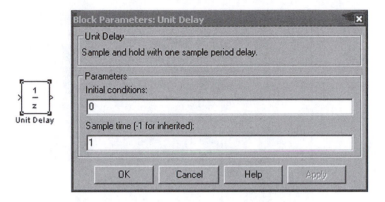

1.6.5.1 Direct Look-Up Table (n-D) Block

This block executes matrix indexing (sampling of individual elements or element groups in a matrix or a vector) (Figure 1.18).

1.6.5.2 Look-Up Table (2-D) Block

This block executes two-dimensional (2-D) table interpolation of data given as the matrix 2-D array (mapping of two input variables — the coordinates of array data elements — into the output variable serving as the array data element) (Figure 1.19).

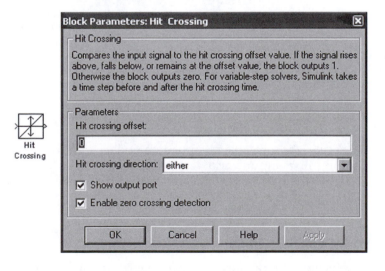

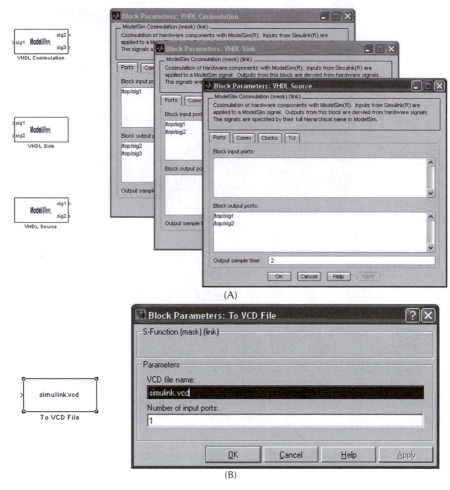

FIGURE 1.17
The VHDL Cosimulation, VHDL Sink, VHDL Source, and To VCD Blocks.

1.6.5.3   Look-Up Table (n-D) Block

This block executes the *n*-D table interpolation of data given as the table *n*-D array (mapping of *n* input variables — the coordinates of array data elements — into the output variable serving as the array data element) (Figure 1.20).

1.6.6   Math Operations Library

This library includes blocks for the performance of various mathematical operations.

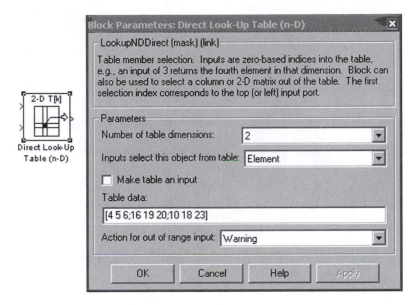

FIGURE 1.18
The Direct Look-Up Table (*n*-D) Block and the Block dialog box.

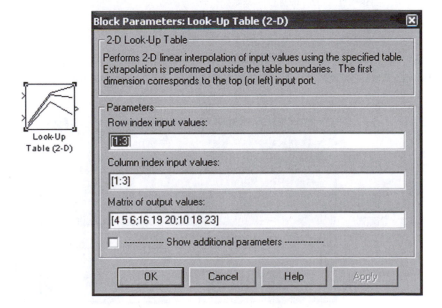

FIGURE 1.19
The Look-Up Table (2-D) Block and the Block dialog box.

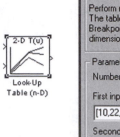

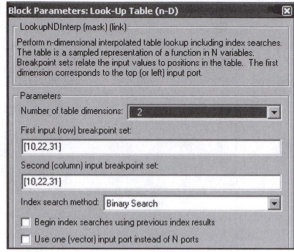

FIGURE 1.20
Look-Up Table (*n*-D) Block and the Block dialog box.

1.6.6.1   Bitwise Logical Operator Block

This block executes bitwise logical operations (AND, OR, XOR, NOT) and shift operations (SHIFT_LEFT, SHIFT_RIGHT) over the integer input variables (sign-free) (Figure 1.21).

1.6.6.2   Combinatorial Logic Block

This block executes the transformation of the input variable (represented by a Boolean-type vector) into the Boolean-type output variable, using the

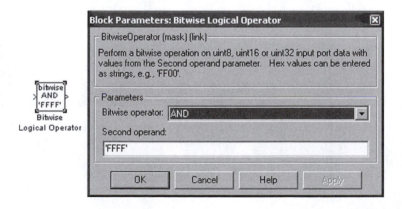

FIGURE 1.21
The Bitwise Logical Operator Block and the Block dialog box.

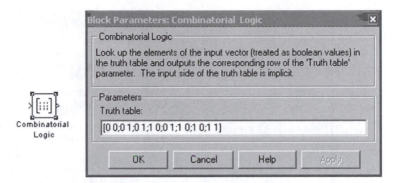

FIGURE 1.22
The Combinatorial Logic Block and the Block dialog box.

matrix assigned as the TruthTable parameter. Each matrix row determines the output variable value for various combinations of component values in input parameter vectors (Figure 1.22).

1.6.6.3 Gain Block

This block executes element-wise multiplication of the input variable (represented by a scalar, a vector, or a matrix) by the constant enhancement factor (Figure 1.23).

1.6.6.4 Logical Operator Block

This block executes the AND, OR, NAND, NOR, XOR, and NOT logical operations by the input variables' processing. Logical operation type and the number of block inputs are relocated parameters (Figure 1.24).

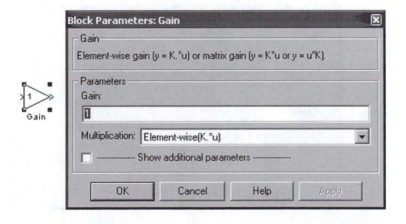

FIGURE 1.23
The Gain Block and the Block dialog box.

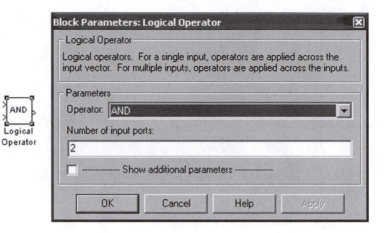

FIGURE 1.24
Logical Operator Block and the Block dialog box.

1.6.6.5   Matrix Gain Block

This block executes matrix multiplication or element-wise multiplication (pairwise product of array elements with conforming dimensionality). The input variable and the enhancement factor that can be defined as scalars, vectors, or matrices are used as operands (Figure 1.25).

1.6.6.6   MinMax Block

This block calculates the maximal or minimal value of the input variable's components. The number of inputs is assigned by a block parameter (Figure 1.26).

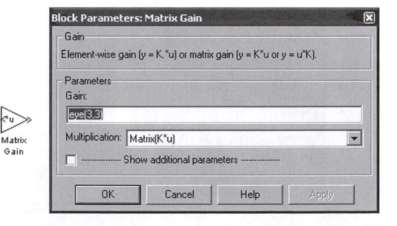

FIGURE 1.25
The Matrix Gain Block and the Block dialog box.

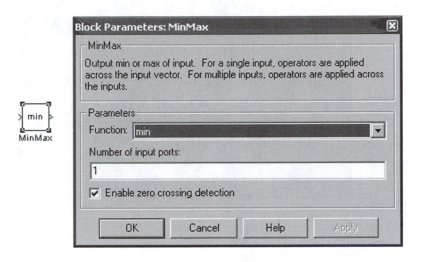

FIGURE 1.26
The MinMax Block and the Block dialog box.

1.6.6.7 *Product Block*

This block executes element-wise multiplication or division of input variables. The number of block inputs and the operation type are assigned as parameters (Figure 1.27).

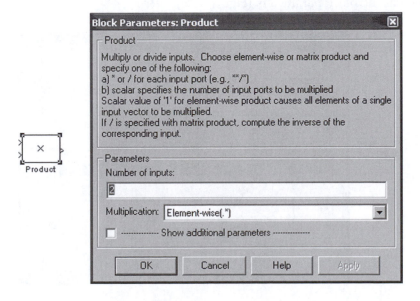

FIGURE 1.27
The Product Block and the Block dialog box.

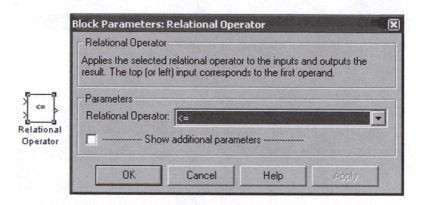

FIGURE 1.28
The Relational Operator Block and the Block dialog box.

1.6.6.8   *Relational Operator Block*

This block executes element-wise comparison of the input variable's components, using relation operations such as =, ≠, <, ≤, ≥, and > (Figure 1.28).

1.6.6.9   *Sum Block*

This block calculates algebraic expressions where input variables participate in the addition and subtraction operations exclusively. The number of inputs and corresponding signs for the addition and subtraction operations are given as the block parameters (Figure 1.29).

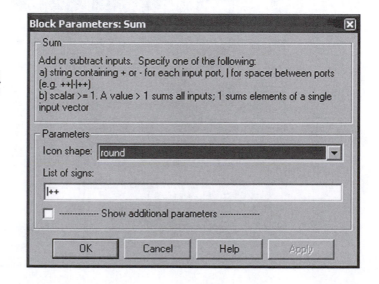

FIGURE 1.29
The Sum Block and the Block dialog box.

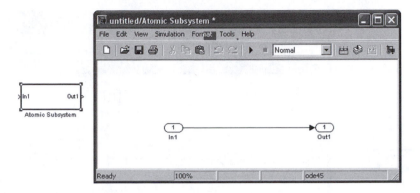

FIGURE 1.30
The Atomic Subsystem Block and the Block window.

1.6.7 Ports&Subsystems Library

This library includes the blocks for subsystem generation, ensuring the Simulink model hierarchy.

1.6.7.1 Atomic Subsystem Block

This block represents a subsystem for a certain system (Figure 1.30).

1.6.7.2 Enabled Subsystem Block

This block represents a subsystem whose execution is possible at a certain resolution input signal value (Figure 1.31).

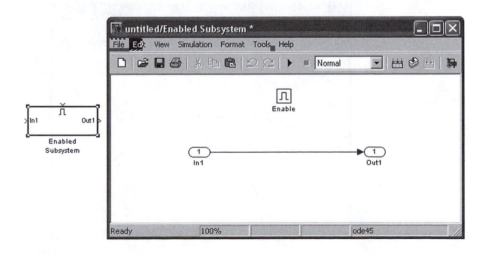

FIGURE 1.31
The Enabled Subsystem Block and the Block window.

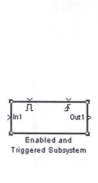

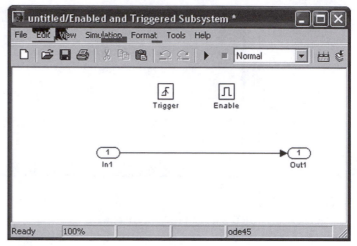

FIGURE 1.32
The Enabled and Triggered Subsystem Block and the Block window.

1.6.7.3   Enabled and Triggered Subsystem Block

This block represents a subsystem whose execution is possible at a certain resolution input signal value or at the flip-flop input activation (Figure 1.32).

1.6.7.4   For Iterator Subsystem Block

This block represents a subsystem activated by each simulation time step (Figure 1.33).

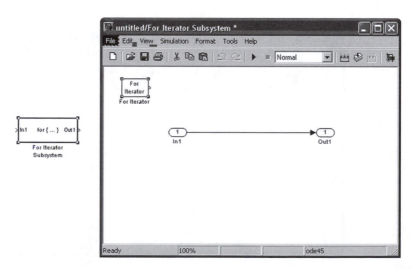

FIGURE 1.33
The For Iterator Subsystem Block and the Block window.

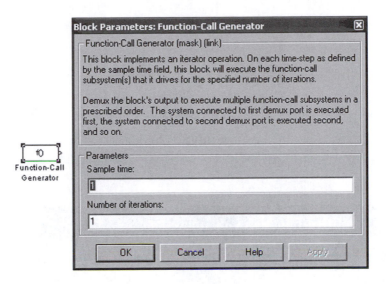

FIGURE 1.34
The Function-Call Generator Block and the Block window.

1.6.7.5 Function-Call Generator Block

This block provides activation of subsystems connected to it with preset periodicity (Figure 1.34).

1.6.7.6 Function-Call Subsystem Block

This block represents a subsystem activated by a given function (Figure 1.35).

1.6.7.7 If Action Subsystem Block

This block represents a subsystem activated by the flip-flop input connected to the If Block output (Figure 1.36).

1.6.7.8 Subsystem Block

This block represents a subsystem of an alternative system (Figure 1.37).

1.6.7.9 Subsystem Examples Block

This block contains the examples of subsystems applications in design (Figure 1.38).

1.6.7.10 Triggered Subsystem Block

This block represents a subsystem whose execution is made possible by the flip-flop input activation (Figure 1.39).

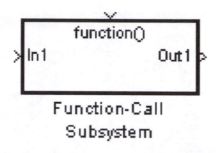

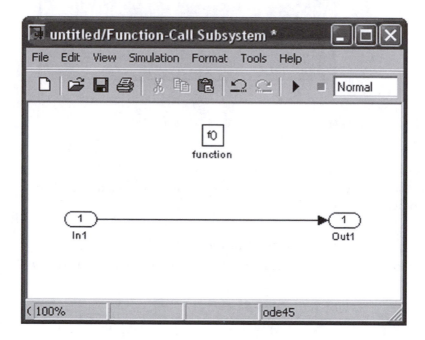

FIGURE 1.35
The Function-Call Subsystem Block and the Block window.

1.6.7.11   Switch Case Block

This block represents a C-like switch-control flow statement (Figure 1.40).

1.6.7.12   Switch Case Action Subsystem Block

This block represents a subsystem whose execution is initiated by the flip-flop input connected to the Switch Case Block output (Figure 1.41).

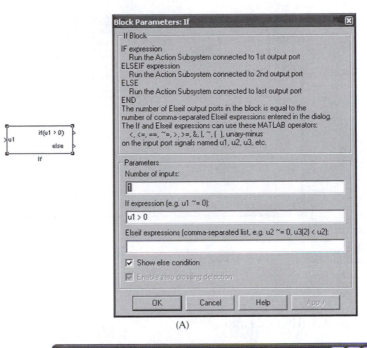

(A)

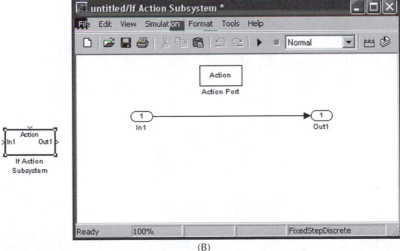

(B)

FIGURE 1.36
The If Action Subsystem Block and the Block windows.

1.6.7.13 *While Iterator Subsystem Block*

This block represents a subsystem whose execution is iterated while a certain condition is met at a given simulation time step (Figure 1.42).

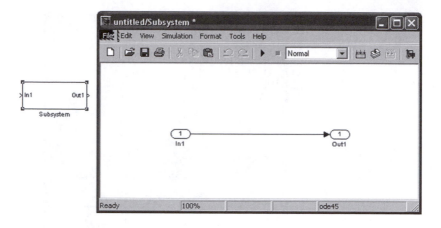

FIGURE 1.37
The Subsystem Block and the Block window.

1.6.8   Signal Attributes Library

This library contains blocks for operation with signal attributes.

1.6.8.1   Data Type Conversion Block

This block transforms the input variable into a given data type (Figure 1.43).

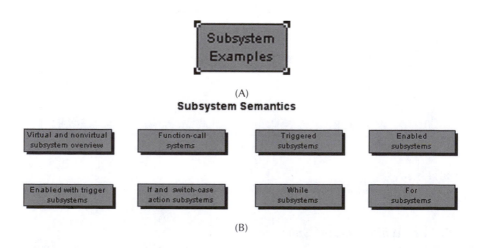

FIGURE 1.38
The Subsystem Examples Block and the Block window.

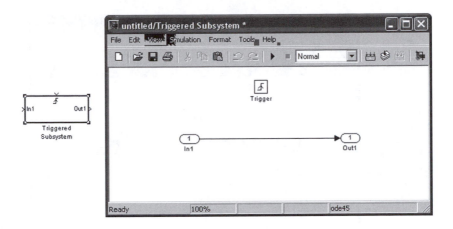

FIGURE 1.39
The Triggered Subsystem Block and the Block window.

1.6.8.2   IC Block

This block assigns a starting value at the block output. The block is expedient to be used for the generation of algebraic circuits that assign the starting approximations (Figure 1.44).

1.6.8.3   Probe Block

This block shapes the output variables whose values correspond to the parameters of a block connected to the given block input (Figure 1.45).

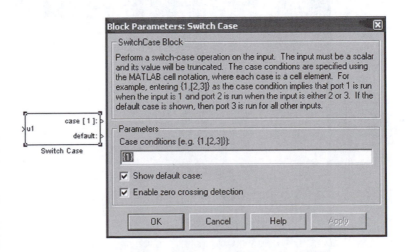

FIGURE 1.40
The Switch Case Block and the Block window.

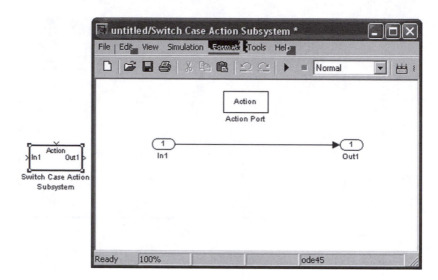

FIGURE 1.41
The Switch Case Action Subsystem Block and the Block window.

1.6.8.4   Rate Transition Block

This block fits the performances of blocks with various operation rates (Figure 1.46).

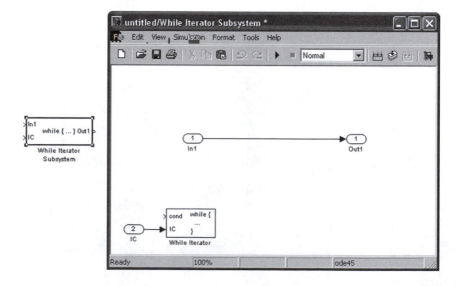

FIGURE 1.42
The While Iterator Subsystem Block and the Block window.

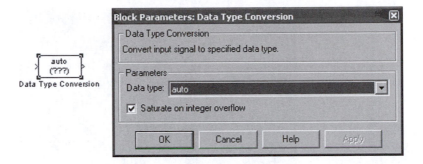

FIGURE 1.43
The Data Type Conversion Block and the Block dialog box.

1.6.8.5   Signal Specification Block

This block defines the block input signal parameters (Figure 1.47).

1.6.8.6   Width Block

This block defines the number of elements in a vector input signal (Figure 1.48).

1.6.9   Signal Routing Library

This library contains blocks for the generation of necessary routes for the model signals.

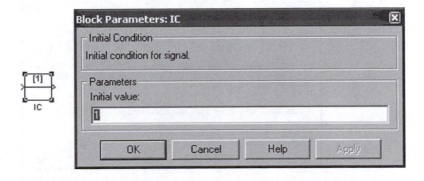

FIGURE 1.44
The IC Block and the Block dialog box.

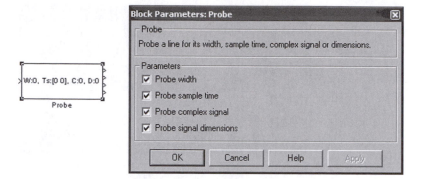

FIGURE 1.45
The Probe Block and the Block dialog box.

1.6.9.1   Bus Creator Block

This block groups signals into a bus for convenient graphic simulation (Figure 1.49).

1.6.9.2   Bus Selector Block

This block brings several input bus-constituting signals to the block outputs (Figure 1.50).

1.6.9.3   Data Store Block

This block stores data at the stored data memory region (Figure 1.51).

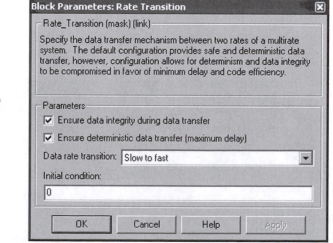

FIGURE 1.46
The Rate Transition Block and the Block dialog box.

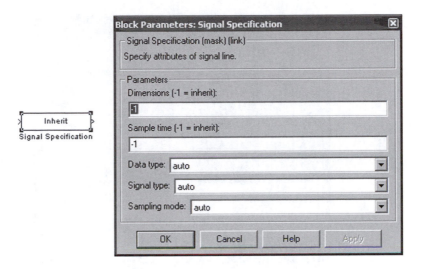

FIGURE 1.47
The Signal Specification Block and the Block dialog box.

1.6.9.4 *Demux Block*

This block extracts scalar, vector, or matrix signals out of the input bus generated by the Mux Block and transfers them to the block outputs. The number of outputs is the block parameter (Figure 1.52).

1.6.9.5 *From Block*

This block operates in conjunction with the Goto Block. The From Block output variable corresponds to the Goto Block output variable. The From Block can be connected only to the Goto Block, whereas the Goto Block can be connected to numerous From Blocks (Figure 1.53).

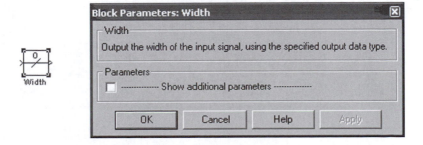

FIGURE 1.48
The Width Block and the Block dialog box.

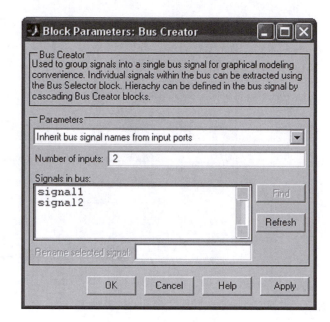

FIGURE 1.49
The Bus Creator Block and the Block dialog box.

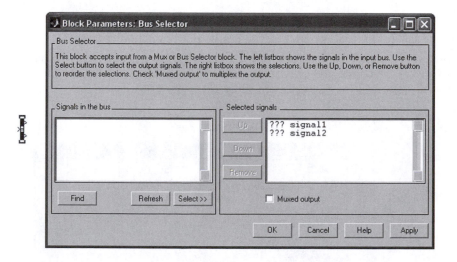

FIGURE 1.50
The Bus Selector Block and the Block dialog box.

FIGURE 1.51
Data Store Blocks.

1.6.9.6   Goto Block

The block input variable is transmitted to all associated From Blocks. Thus, the employment of the two blocks — From and Goto — make possible the transmission of signals among the blocks without signal line outlining for convenient graphic simulation (Figure 1.54).

1.6.9.7   Manual Switch Block

This block allows switching between two input ports when the left mouse button is double-clicked on the block image (Figure 1.55).

1.6.9.8   Multiport Switch Block

This block transmits output data from the input whose number is assigned at the block first input. The number of block inputs is its parameter (Figure 1.56).

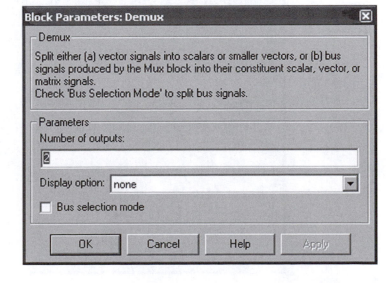

FIGURE 1.52
The Demux Block and the Block dialog box.

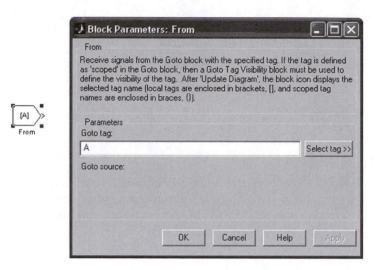

FIGURE 1.53
The From Block and the Block dialog box.

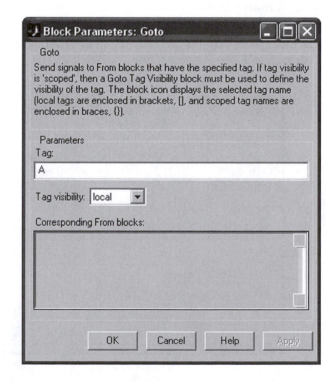

FIGURE 1.54
The Goto Block and the Block dialog box.

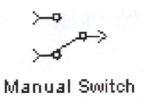

Manual Switch

FIGURE 1.55
The Manual Switch Block.

1.6.9.9   Mux Block

This block integrates the block input scalar, vector, or matrix signals into the output bus. The number of block inputs is its parameter (Figure 1.57).

1.6.9.10   Selector Block

This block extracts a certain data element out of the input variable (matrix or vector) in accordance with the given coordinates (Figure 1.58).

1.6.9.11   Switch Block

This block transmits the first input signal to the output if the second input controlling signal satisfies the assigned criterion; otherwise, the third input data are transmitted to the output. The criterion type and the Threshold value are the block parameters (Figure 1.59).

1.6.10   Simulink Extras Library

1.6.10.1   Clock Block

The Clock Block is the source of pulses for flip-flop synchronization (Figure 1.60).

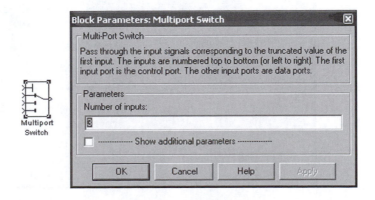

FIGURE 1.56
The Multiport Switch Block and the Block dialog box.

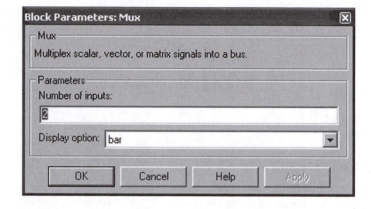

FIGURE 1.57
The Mux Block and the Block dialog box.

1.6.10.2   DFF Block

This is the DFF model (Figure 1.61).

1.6.10.3   D Latch Block

This is the D Latch model (Figure 1.62).

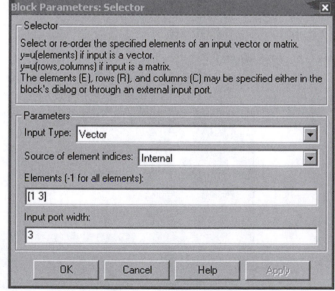

FIGURE 1.58
The Selector Block and the Block dialog box.

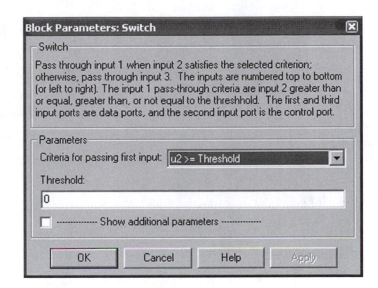

FIGURE 1.59
The Switch Block and the Block dialog box.

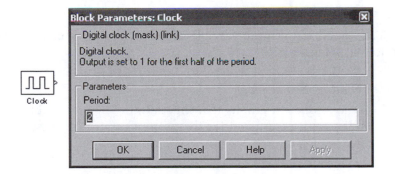

FIGURE 1.60
The Clock Block and the Block dialog box.

FIGURE 1.61
The DFF Block and the Block dialog box.

FIGURE 1.62
The D Latch Block and the Block dialog box.

1.6.10.4 JKFF Block

This block is the JKFF model (Figure 1.63).

1.6.10.5 SRFF Block

This is the SRFF model (Figure 1.64).

1.6.11 Sinks Library

The Sinks Library contains the simulation result mapping blocks.

1.6.11.1 Display Block

This block visualizes the input variable values in a given number representation format (Figure 1.65).

1.6.11.2 Floating Scope Block

This block plots function graphs that illustrate the input variables versus time dependences. It is destined for the prompt connection to signal lines (Figure 1.66).

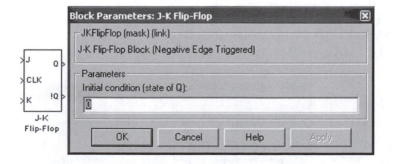

FIGURE 1.63
The JKFF Block and the Block dialog box.

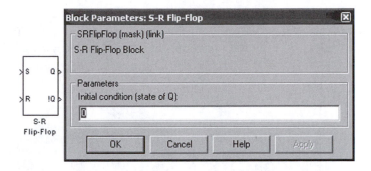

FIGURE 1.64
The SRFF Block and the Block dialog box.

1.6.11.3 Out1 Block

This block generates output ports for the model subsystems (Figure 1.67).

1.6.11.4 Scope Block

This block displays the block input variables versus time graphs (Figure 1.68).

1.6.11.5 Stop Simulation Block

This block stops simulation in response to the "input signal value differs from zero" event (Figure 1.69).

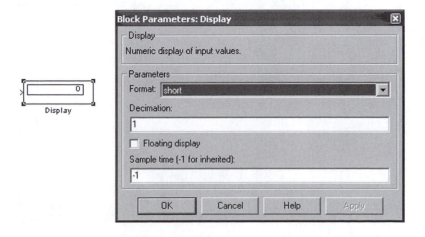

FIGURE 1.65
Display Block and the Block dialog box.

1.6.11.6   Terminator Block

This block is meant for connection to the block-free output port and correction of the fault caused by the nonconnected signal line (Figure 1.70).

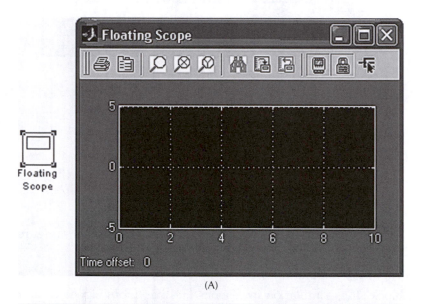

(A)

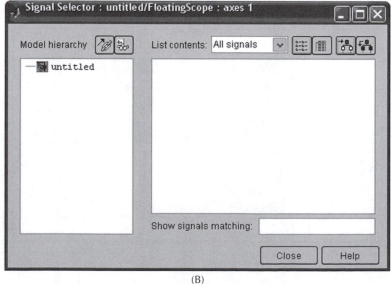

(B)

FIGURE 1.66
The Floating Scope Block and the Block dialog box.

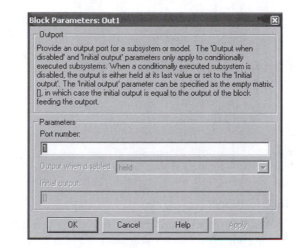

FIGURE 1.67
The Out1 Block and the Block dialog box.

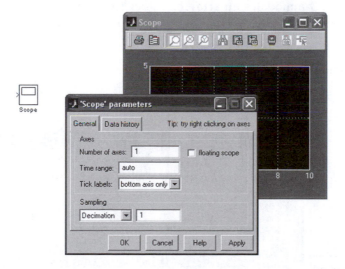

FIGURE 1.68
The Scope Block and the Block dialog box.

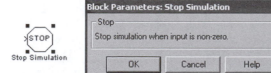

FIGURE 1.69
The Stop Simulation Block and the Block dialog box.

FIGURE 1.70
The Terminator Block and the Block dialog box.

1.6.11.7   To File Block

This block stores the input data array (including the time point vector) in the MATLAB file with the *.mat extension (Figure 1.71).

1.6.11.8   To Workspace Block

This block accommodates an array of values (or a structure) assigned to a concrete variable in the MATLAB workspace (Figure 1.72).

1.6.12   Sources Library

The Sources Library contains blocks for the assignment of input impacts.

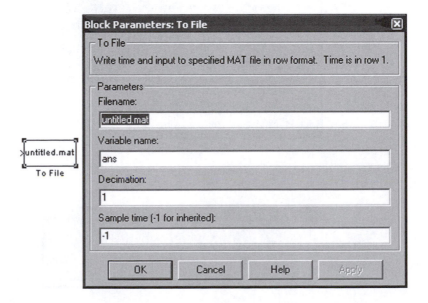

FIGURE 1.71
The To File Block and the Block dialog box.

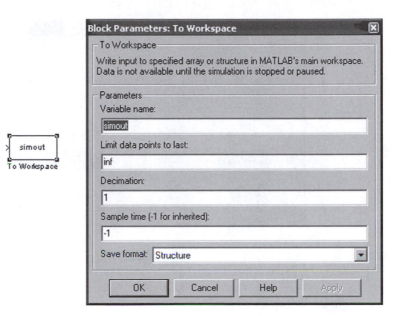

FIGURE 1.72
The To Workspace Block and the Block dialog box.

1.6.12.1   Constant Block

The output variable is assigned some constant value (a vector or a scalar) (Figure 1.73).

1.6.12.2   Digital Clock Block

The output variable is assigned the current simulation time value (the sampling is made with a preset period) (Figure 1.74).

1.6.12.3   In1 Block

This block generates an input port for the model subsystem (Figure 1.75).

1.6.12.4   From File Block

This block generates a signal from the array that has the time point vector as its first row and the corresponding variable vectors as its subsequent rows (Figure 1.76).

1.6.12.5   From Workspace Block

The input impact is generated by an array or a structure accommodated in the MATLAB workspace (Figure 1.77).

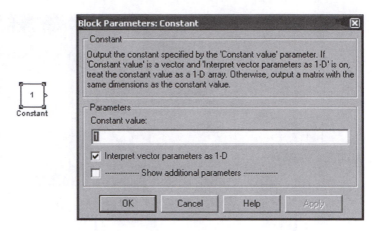

FIGURE 1.73
The Constant Block and the Block dialog box.

1.6.12.6 Pulse Generator Block

This block shapes the input impact as a sequence of square pulses. The block parameters are its amplitude, phase, and pulse length (Figure 1.78).

1.6.12.7 Signal Builder Block

This blocks permits the generation of piece-wise linear signal sources and their employment as the model input impacts (Figure 1.79).

1.6.12.8 Signal Generator Block

This block generates a periodic signal such as sinusoid, square pulses, saw-tooth pulses, or a random signal. Its parameters are the signal amplitude and frequency (Figure 1.80).

FIGURE 1.74
The Digital Clock Block and the Block dialog box.

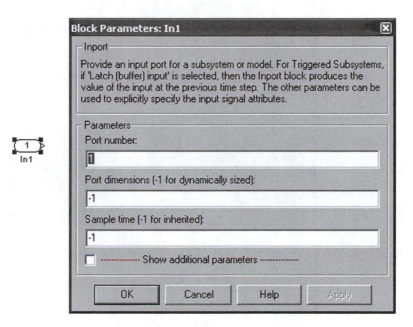

FIGURE 1.75
The In1 Block and the Block dialog box.

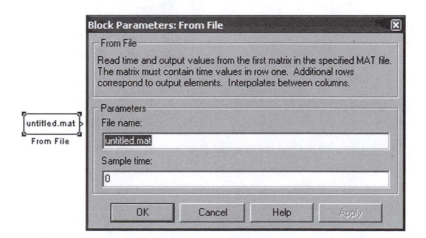

FIGURE 1.76
The From File Block and the Block dialog box.

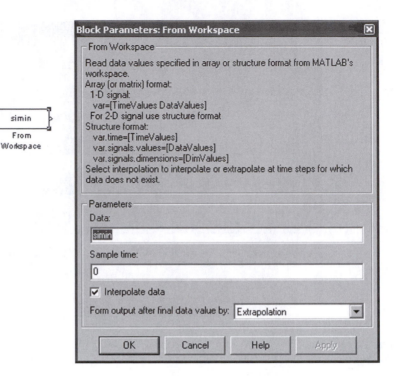

FIGURE 1.77
The From Workspace Block and the Block dialog box.

1.6.12.9 Step Block

This block generates input impact as a stepwise function. Its amplitude, jump time, and starting value are the block parameters (Figure 1.81).

1.6.13   Stateflow Library

The Stateflow Library contains only one Chart Block for model generation as finite state machine (FSM) state diagrams (see Chapter 2).

1.6.13.1 Chart Block

This block contains a Stateflow diagram constructed by the user from Stateflow Chart Blocks (Figure 1.82). The Stateflow Component (event-driven simulation), equipped with its own graphical user interface (see Chapter 2), is accessible to users only via Simulink, and the models generated with its help are incorporated into Simulink models as Chart Blocks.

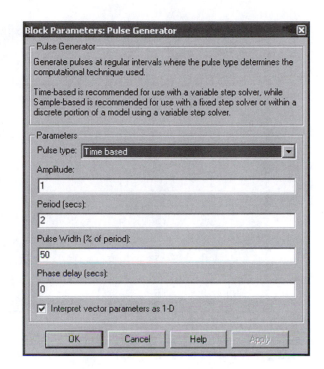

Pulse
Generator

FIGURE 1.78
The Pulse Generator Block and the Block dialog box.

1.6.14　User-Defined Functions Library

1.6.14.1　Fcn Block

The Fcn Block permits the user to apply the expression written in C language style to the input (Figure 1.83).

1.6.14.2　MATLAB Fcn Block

This block permits the application of a MATLAB function to the input (Figure 1.84).

1.6.14.3　S-Function Block

This block provides access to S-functions from the block diagram. The S-function can be written in C language, Fortran, Ada, or MATLAB M language, according to definite rules (Figure 1.85).

1.6.14.4　S-Function Builder Block

This block Creates the S-function in C language, using the specifications and the user's starting C code (Figure 1.86).

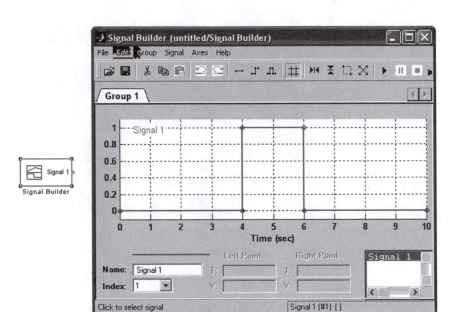

FIGURE 1.79
The Signal Builder Block and the Block dialog box.

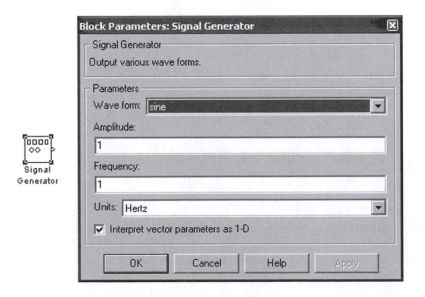

FIGURE 1.80
The Signal Generator Block and the Block dialog box.

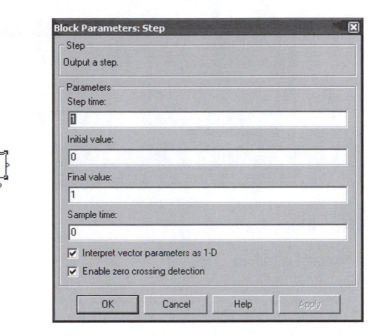

FIGURE 1.81
The Step Block and the Block dialog box.

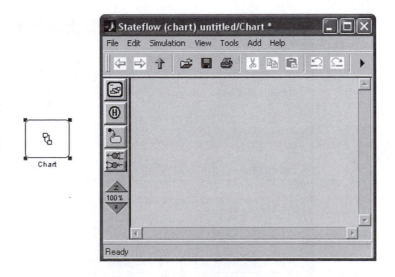

FIGURE 1.82
The Chart Block and the Block window.

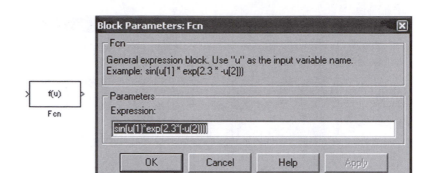

FIGURE 1.83
The Fcn Block and the Block dialog box.

1.6.15   Additional Libraries

In addition to the basic blockset libraries, some additional libraries and utilities outside the Simulink core can also be used for design of test benches and DUT models. The data on such facilities are given in the MATLAB Help window. We present here most of these libraries and utilities:

1. Logic and Bit Operations Library
2. Model Verification Library
3. Additional Math&Discrete Library

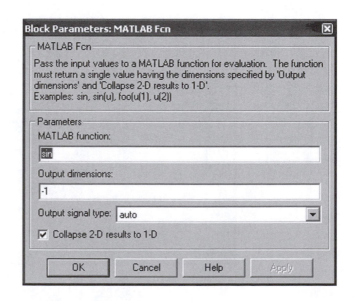

FIGURE 1.84
The MATLAB Fcn Block and the Block dialog box.

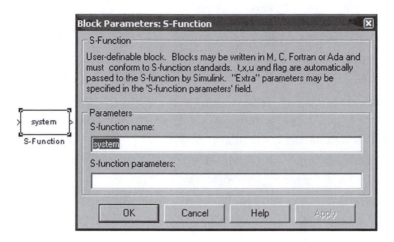

FIGURE 1.85
The S-Function Block and the Block dialog box.

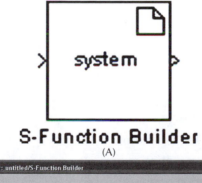

(A)

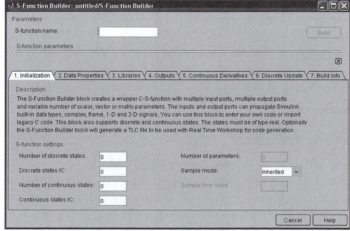

(B)

FIGURE 1.86
The S-Function Builder Block and the Block dialog box.

4. Aerospace Blockset

5. CDMA Reference Blockset

6. Communications Blockset

7. Control System Toolbox

8. Dials&Gaudes Blockset

9. Embedded Target for Infineon C166® Microcontrollers

10. Embedded Target for Motorola® HC12

11. Embedded Target for Motorola® MPC555

12. Embedded Target for OSEK/VDX

13. Embedded Target for TI C2000 DSP

14. Embedded Target for TI C6000 DSP

15. Fuzzy Logic Toolbox

16. Model Predictive Control Toolbox

17. Neural Network Blockset

18. RF Blockset

19. Real-Time Workshop Embedded Coder

20. Report Generator

21. Signal Processing Blockset

22. SimMechanics

23. SimPowerSystems

24. Simulink Control Design

25. Simulink Parameter Estimation

26. Simulink Response Optimization

27. System Identification Toolbox

28. Virtual Reality Toolbox

29. XPC Target

1.7 Simulink Debugger

The interactive debugger is a program product devised for fault finding and for the control of program code execution and its interruption in case of warning or fault, as well as for the control of point setting. Debugging permits the user to control the variable values during the program execution. Simulink Debugger ensures that a number of possibilities are implemented in typical programming systems. Simulink Debugger aids in setting the debugging conditions for various block diagram elements, in stepwise debugging, and in interruption determining based on a given criterion [1,2].

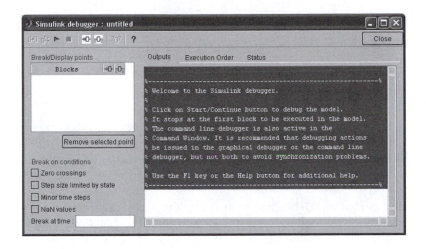

FIGURE 1.87
The Simulink Debugger window.

Simulink Debugger is started from the Debugger dialog window or from the MATLAB command line. To debug a model, the model window should be opened and the debugging mode started by Simulink Debugger instruction in the Tools menu or by clicking the mouse cursor on the Debug icon in the model window toolbar. The Simulink Debugger window is shown in Figure 1.87.

The debugger toolbar has eight icons devised for setting various debugging conditions and for the debugging control:

1. Step to Next Nlock
2. Go to Start of Next Time Step
3. Start/Continue
4. Stop Debugging
5. Break before Selected Block
6. Display I/O of Selected Block when Executed
7. Display Current I/O of Selected Block
8. Help

The dialog window is closed with the Close button. When the debugging mode is started, the model window expands for browser region mapping. The Outputs panel shows debugging mode and switches on the tabulation positions for mapping blocks operation sequence and the debugger state. The Break/Display Points group of options has a list of blocks, for which the simulation halt points or condition checkpoints are determined. The Break on Conditions group of options includes a set of checkboxes that

```
[Tm = 0                           ] **Start** of system 'ASM' outputs

(sldebug @0:0 'ASM/SubSystem/Step'):
```

FIGURE 1.88
A Debugging message.

control the simulation-related condition checkpoints and a text box for the
input of halt points by time. To start debugging, the Start/Continue icon
should be clicked. When this is done, the first block to be executed is
indicated by color, and a confirming message appears at the debugger
panel (Figure 1.88).

During Simulink Debugger operation, the user can interact with the
debugger by using toolbar icons or by the input of instructions in the MAT-
LAB window and selection of corresponding blocks using the mouse in the
model window.

Simulink returns the list of all model-included blocks sorted according to
their execution order and assigned as the s:b structure, where s is a subsystem
number, b is a block number in the subsystem, and the enumeration is started
from zero. The Execution Order panel ensures the possibility of viewing the
complete expanded list of blocks. Many Simulink commands introduced
from the MATLAB command line use block index as a parameter value.
Simulink permits the user to determine the block for these commands in
two different ways: by setting the block index or by using the gcb command
(with regard to the block chosen for a given moment) [1,2].

Model debugging in Simulink stipulates simulation performance as a sin-
gle-step stepwise calculation procedure or is performed by transition from
block to block using block sorting. To employ this mode, the Step to Next
Block icon should be clicked at each step. By model debugging from block
to block, a message is displayed for the block being studied, and it is indicated by color in the model window. The message includes block index,
block number, and the values of input and output variables. To employ the
stepwise mode, the Go to Start of Next Time Step icon should be clicked at
each step [1,2].

In some cases simulation should be halted because of a fault condition.
The debugger commands determine debugging modes and corresponding
parameters called the checkpoint. Simulink debugger includes a number of
conditions associated with warnings or faults. Debugger causes the simulation to halt when the condition checkpoints are set as:

1. The specified block is achieved.
2. The specified simulation time is achieved when a specified event
   appears.

To proceed with the simulation up to the next checkpoint (or up to the
process termination if halt points are absent), the Start/Continue icon should

be clicked in the debugger dialog window. To proceed with the simulation, ignoring all checkpoints, the Run instruction in the MATLAB command line should be used [1].

To set the block debugging mode and checkpoint, a block should be selected and the Break before Selected Block icon clicked by the mouse. After that, the line that corresponds to the selected checkpoint with the determined halt point checkbox is added in the list of the Break/Display Points group of options.

The Simulink Debugger function can result in simulation interrupt based on the time halt checkpoint. One value of time halt can be given by assigning the desirable time value in the Break at Time text field. The required sequence of time halt checkpoints is set by the Tbreak Time instruction. After that the Start/Continue icon is employed to stop the simulation. To delete the current halt checkpoint, the Tbreak instruction is used.

Debugger can result in simulation interrupt if it discovers events that can lead to calculation errors or to an integration rate decrease. The corresponding checkpoints are set in the Break on Conditions group of options in the debugger window, using checkboxes. Selecting the Zero Crossing checkbox interrupts task execution if the zero crossing phenomenon is revealed. The Step Size Limited by State checkbox is intended for finding the system state variables related to the decrease in simulation rate. Selecting the NaN Values checkbox halts the simulation, when any signal value that cannot be represented as a number reaches the figure. Complete information on Simulink Debugger commands is given in [1,2].

Statistical analysis of operation results for test benches created as Simulink models can be performed by means of the MATLAB library of functions, Statistics Toolbox.

Despite the extensive functional potential of Simulink Block Libraries, Simulink is, nevertheless, just a MATLAB component and can operate only under its control. To overcome the inseparability of a Simulink-generated model and the development environment, Simulink incorporates the tool called Real-Time Workshop (RTW). This tool provides for the creation of Simulink model–based software (in our context, test benches) intended for the real-time control of target-specific hardware (in our context, ATE [automated test equipment]). RTW's major functional features are:

1. The ability to generate program code from any Simulink model (the only limitation is induced by MATLAB Fcn Block and S-function Block) that should be preliminarily converted into MEX-files (Figure 1.89) [5]

2. Use of an expandable library of device drivers

3. An automatic and fully customized software development process

4. Unification of components controlled by various operating systems

5. Program code generation for any standard (conforming to the ANSI requirements) C-Language compiler

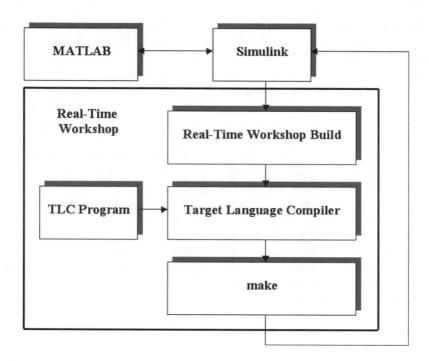

FIGURE 1.89
RTW's open architecture.

RTW can assist in solving the following tasks:

1. Software development for real-time control systems. The system's logical structure is determined with the aid of MATLAB and Simulink, and then the program code is generated directly from the system's Simulink model. The program obtained can be compiled and loaded directly to the target hardware.

2. Creation of real-time signal processing algorithms. The algorithm is developed by the MATLAB and Simulink facilities, and then the program code is generated and customized to the specific hardware.

3. Creation of test benches that include both Simulink models and real target hardware. Using such test benches, the behaviors of real systems (DUT) can be studied in varying conditions, and their reliability and productivity with various input impact values can be assessed.

4. Real-time interactive adjustment of systems parameters. In this case Simulink is used as a connecting link between the researcher and real-time software (test), permitting changes to the parameters of

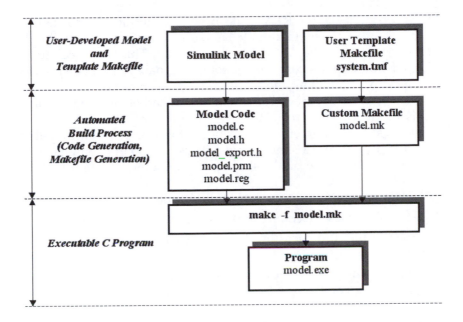

FIGURE 1.90
The Build process.

testing the hardware–software system (test bench) without restarting the test.

5. Generation of transferable C-code for export to alternative real-time systems.

6. An enhanced rate of model experiment conduct (in our context, test development and debugging).

Creation of independent RTW code is fully automated and based on the consecutive performance of three key stages by the user (Figure 1.89 and Figure 1.90) [5]:

1. User-developed model and template makefile
2. Automated build process
3. Executable C program

At the first stage two files are prepared as follows:

1. **Simulink model file:** The mdl-file (the Simulink model-containing file)
2. **Template makefile:** The tmf-file (on its base, the RTW Build process and make utility create the required version of real-time executable C-code: for instance, for the preparation of Embedded C for Visual C/C++, the tmf-file is the ert_vc.tmf; a user can control the RTW performance by modifying the tmf-file [5])

At the second stage the RTW Build process controlled by m-file under the name of make_rtw.m (Figure 1.90) is executed and results in a Simulink model description in the ASCII (rtw-file) format. This intermediate Simulink model description contains the following data in particular: parameter values for a Simulink model, vector signal dimensionality, discrete changes in model time, and so on. After that, the Target Language Compiler (TLC) is activated, which reads the rtw-file and using the TLC Program (a set of tlc-files: system target file, block target file, and TLC function library [Figure 1.89]), generates a Simulink model description in C Language. TLC generates five files based on the starting Simulink model (Figure 1.89) [6]:

1. **c-file:** Contains C-code for the Simulink model
2. **h-file:** Describes the connections among the model blocks
3. **export h-file:** Contains the data on exported signals and model parameters
4. **prm-file:** Contains the parameters of model blocks
5. **reg-file:** Contains the parameters of the model modeling time

At the third stage the make utility generates executable C-code, that is, the real-time exe-file. The exe-file is generated from the above-mentioned files and from the Custom Makefile, the mk-file, generated previously on the basis of the tmf-file. At this stage the S-function codes are added to the generated file if they are found in the Simulink model. If the S-function is written in C language or in FORTRAN, it is preliminarily compiled and converted into a MEX-file. The generated MEX-file is dynamically connected to alternative model components whenever necessary.

The generation algorithm for executable C-code (RTW Control Logic) is given in Figure 1.91. The specific work sequence of RTW operation is described in [5].

The exe-file generated by the make utility can be loaded to target specific hardware. When Simulink is started in external mode (Simulink External Mode), final parameters for the generated software can be adjusted directly during its operation. In MATLAB-based Simulink External Mode, a new program code development environment is created (Rapid Prototyping Program Framework). When operating in this environment, a user can modify block parameters of the starting Simulink model and change the list of model input and output parameters without recompiling the RTW code. Interaction of software programs in this environment is based on the client–server calculation model, wherein Simulink is a client sending queries to the server (external program) to determine new values of a model parameter. The environment structural organization makes it an expandable system and permits application of various interaction protocols for its components.

After the RTW code is started, Simulink sets initial block parameters, executes the Simulink model update command, and passes into an expectation state, remaining in this state until the user changes the Simulink model parameters or until an external program message arrives. After that, new parameter

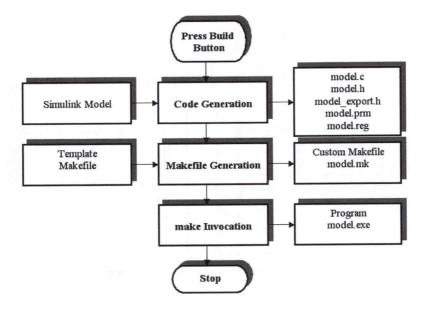

FIGURE 1.91
The logic that controls automatic program building.

values are sent to the executable external program. Parameter values are transferred as the arguments of a MEX-file that is called dynamically by Simulink. The MEX-file has a code that implements one side of the interprocess communications (IPC) channel. The channel connects the Simulink process (in whose framework the MEX-file is executed) with the external program-executing process. Via this channel, the MEX-file transmits new parameter values to the external program that records new parameters in the structural variable, SimStruct (data structure that contains all data related to the model code). These two interprocesses can be remote (executed by different devices) — in this case the communication protocol is used for data transfer — or local — the shared memory is used for data transfer (Figure 1.92) [7].

With external mode operation the Simulink model-determining parameters cannot be changed. These include:

1. The number of states, inputs, and outputs for any block
2. The sample time value
3. The integration (calculation of states) algorithm for continuous systems
4. Names of the model or any of its blocks
5. The Fcn Block parameters

The specific work sequence of the Simulink External Mode operation is described in [7].

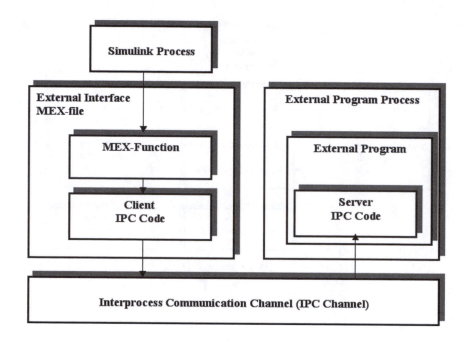

FIGURE 1.92
External Mode architecture.

References

1. Using Simulink: Dynamic System Simulation for MATLAB. The MathWorks, Inc., Natick, MA, 2004.
2. Dabney J.B., Harman T.L. Mastering Simulink 4 (2nd ed.). Prentice Hall, New York, 2001.
3. Stateflow User's Guide. The MathWorks, Inc., Natick, MA, 2004.
4. Link for ModelSim for Use with MATLAB and Simulink User's Guide. The MathWorks, Inc., Natick, MA, 2004.
5. Real-Time Workshop: For Use with Simulink. User's Guide. The MathWorks, Inc., Natick, MA, 2004.
6. Using Simulink: Dynamic System Simulation for MATLAB. Target Language Compiler Reference Guide. The MathWorks, Inc., Natick, MA, 2004.
7. Real-Time Windows Target: For Use with Real-Time Workshop. User's Guide. The MathWorks, Inc., Natick, MA, 2004.

2

Stateflow®: Creating Finite State Machine Models

2.1 Introduction

Stateflow software is the extension of Simulink® and serves as a tool for model building and event-driven hybrid system simulation based on Harel's Statecharts [1–5]. In Harel's formalism the discrete variable $q = \{q_1,..., q_k\}$ finds congruence in $Q = \{Q_1,..., Q_k\}$ set of graph nodes (states), one of them being marked as a starting node. Graph nodes are joined by $\{Q_i, Q_j\}$ arcs demonstrating what new state can be attained by the statechart at the advent of some preset events. In other words, the statechart key elements are the States and the Transitions. The events are described by predicates over a set of variables, $X = \{x_1,..., x_m\}$, whose true values are treated as the transition functioning conditions. The predicates are called the state change conditions. If $X = \{x_1,..., x_m\}$ are continuous variables describing the behavior of the q parameter-dependent dynamical system, then the predicates isolate specific events in its phase space. The statechart operates in continuous time and is the dynamic-system description element.

Two additional mechanisms are introduced for the convenient description of sophistic behavioral mechanisms: (a) the mechanism for generation of hierarchical statecharts and (b) parallel operation of several states.

Embedding implies the recursive description technique for the statecharts, suggesting that each graph node can, in its turn, be represented by a statechart. Such a node is called the hypernode. It corresponds to the superstate (hyperstate) that unites several states with identical responses to the same event. Transition to the hypernode is treated as the transition to the starting node of an embedded statechart. The *historical pseudo state* (denoted by the letter H) is used here. When the historical pseudo state is employed, returning to the hyperstate conveys the control to that system state in which it last resided before leaving the given hyperstate. In such a model the interstate transitions are caused either by conditions (advent of the predicate truth above the statechart's internal variables) or by external events.

The second mechanism provides the possibility of being in several inter-connected "parallel states" at the same time, thus permitting simultaneous control of different independent processes.

In this manner for the description of the set of successive different behaviors, Harel's formalism uses:

1. An oriented graph with two types of nodes: simple nodes and hypernodes, some of them being marked as the starting ones
2. A set of discrete variables with preset values assigned to the graph nodes
3. A set of discrete or continuous variables describing a concrete discrete or continuous behavior of a dynamic system
4. A set of predicates over the set of discrete or continuous variables used in defining the transition conditions to a new local behavior of a dynamic system
5. A set of algorithms ascribed to arcs for a behavior initiation by transition to the new state

Statecharts are quite beneficial for the development of discrete systems and are used at present for hybrid system design as well. To this end, separate descriptions of discrete and continuous behaviors and application of state-charts for the continuous processes control are employed.

Simulink models frequently incorporate the subsystems in the shape of finite automata (systems with discrete time, a finite set of internal states, input and output alphabets, output message generation rules, and transitions between the internal states). The finite state machine (FSM) is conventionally represented by state diagrams. A state diagram is composed of the images of states interconnected by the branches that reflect the interstate transitions and the conditions for their execution as well as the output signals related either to the internal states (Moore automata) or to the interstate transitions (Mealy automata).

The Stateflow diagram is the advanced state diagram. It provides graphic representation of internal states, denoting its starting condition and the events described by predicates over the set of discrete or continuous variables.

If a predicate assumes the true value, the corresponding transition between two states is executed. The Stateflow diagram also permits the fanout, using the composite transitions that include the consecutive events and the decision making points. In addition, the Stateflow diagram enables the existence of a diagram hierarchy (each internal state cab, in its turn, is represented by a Stateflow diagram) and parallel execution of several states [3,5].

System model generation by Simulink and Stateflow facilities includes the following stages [3,5]:

1. Creation of the model *combinational* part in Simulink
2. Creation of the model *memory* as FSM in Stateflow
3. Addition of an event and data interface to the Stateflow chart block
4. Model debugging and code obtaining

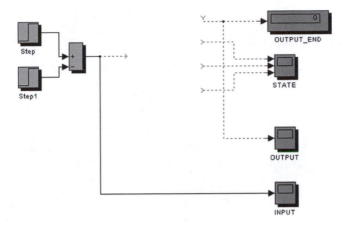

FIGURE 2.1
Generation of a model combinational part in Simulink.

2.2 Creating Charts

For example, we can create a simple model in Simulink, which in turn uses the Stateflow chart block. The generation procedure has the following stages.

1. A new model is created in Simulink (Figure 2.1).
2. Parameters for model blocks are assigned (see Chapter 1).
3. The Chart Block is transferred from the Stateflow blockset library (Figure 2.2).

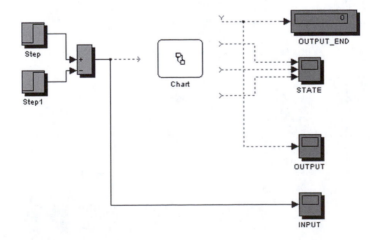

FIGURE 2.2
Dragging the Chart Block into the Simulink model.

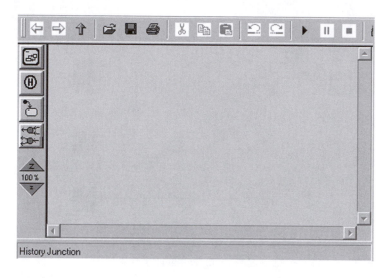

FIGURE 2.3
The Stateflow window.

4. A Stateflow window is called by double-clicking the left mouse button on the Chart Block (Figure 2.3).

5. The FSM internal states are placed in the model window (Figure 2.4).

6. Transitions are formed between the internal states (Figure 2.5).

7. The starting (initial) state is labeled (Figure 2.6).

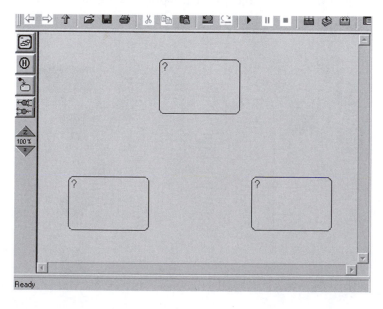

FIGURE 2.4
Placement of the Stateflow diagram internal states in the Stateflow window.

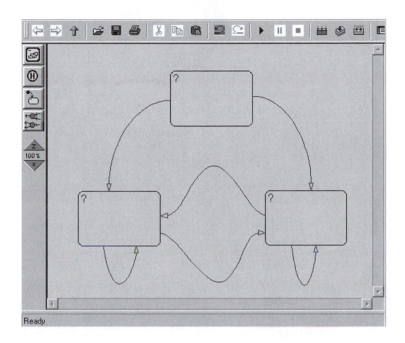

FIGURE 2.5
Generation of transitions between the Stateflow diagram internal states.

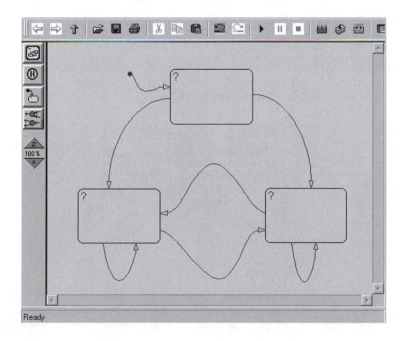

FIGURE 2.6
Labeling of the Stateflow diagram starting (initial) state.

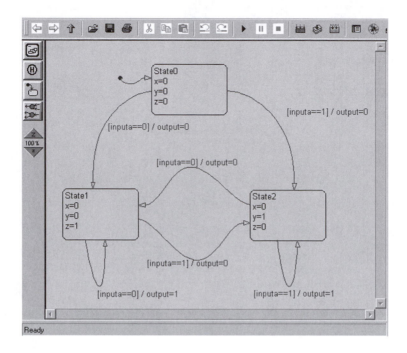

FIGURE 2.7
Label add-in to the Stateflow diagram transitions and internal states.

8. Transitions and internal states are labeled (Figure 2.7).

9. Data input and output ports are set (Figure 2.8).

10. Simulation parameters are set (see Chapter 1) for the model shown in Figure 2.9 and the execution is followed by graphically represented results, using Scope Blocks (Figure 2.10).

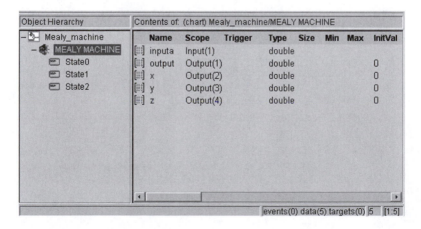

FIGURE 2.8
Input and output data ports for the Stateflow diagram.

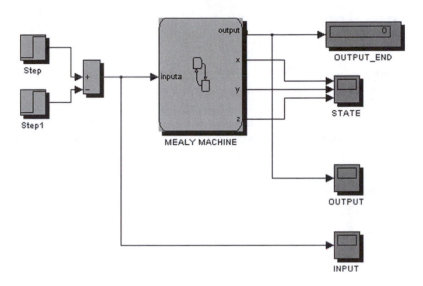

FIGURE 2.9
A model to be simulated in Simulink environment.

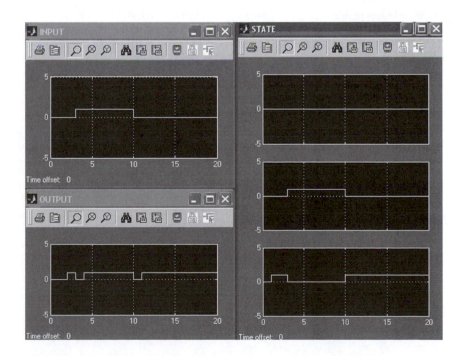

FIGURE 2.10
Graphical representation of simulation results.

2.3   Entering a Stateflow Diagram

A Stateflow diagram contains two basic groups of elements: graphic and nongraphic. The Stateflow diagram graphic part includes states (hyperstates), transitions, default transitions, history junctions, and connective junctions, represented by corresponding icons in the Stateflow window (in Figure 2.3 see the lower screen prompt, where the mouse pointer is above a certain icon):

1. State
2. Default Transition
3. Connective Junction
4. History Junction

Yet another graphic element, Transition, is assigned implicitly.

2.3.1   Internal States

The states can be simple and composite (hyperstates, i.e., the states with hierarchical structure). The states can be united into the chains of states that operate in parallel (and independently of each other) and in succession (Figure 2.11). Each state describes one operation mode of an event-driven system. The state becomes active if it assumes the TRUE value of the transition condition (the transition activates), resulting in this state, or if the state

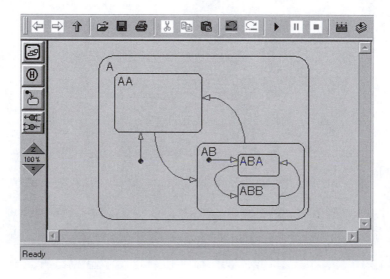

FIGURE 2.11
Stateflow diagram hyperstates.

is the starting state. In Stateflow diagram each state has a *parent*, and by default the parent is the Stateflow diagram by itself.

Internal states and hyperstates are depicted as the rectangles with rounded angles that are called by the State icon. To change the state sizes, a corresponding rectangle should be selected and its required side or face should be moved by the mouse.

For hyperstate creation, a state image should be placed in the Stateflow window and its size should be magnified so that its drawing encompasses all internal states incorporated in the given hyperstate. An alternative method is the generation of a hyperstate drawing with subsequent placement of internal state images inside it. The state embedding may be arbitrary. The state can be transferred from a parent hyperstate by grasping it with the mouse and dragging it to the new parent state.

A hyperstate with all its states can be grouped into a single whole. To do this, the right mouse button should be clicked and the Make Contents/Group instruction chosen in the window (or double-click the left mouse button in the hyperstate box). The grouped states are shown by the thick line. The states can be ungrouped again by double-clicking the left mouse button in the hyperstate box.

To delete an internal state, the corresponding image should be chosen and the Delete key pressed.

A state has text labels that determine the actions executed during its functioning. A state name is first introduced. In describing the state, the following actions can be defined:

1. **Entry:** The action is executed by entry to the state.
2. **During:** The action executed when the state is active.
3. **Exit:** The action executed by exiting from the state.
4. **On: <event_name>:** The action is executed by the event advent, provided that the system is in this state.

The states found inside the hyperstate can be activated if the parent state is active.

The parallel-state system admits the concurrent existence of several active states so that each of them operates independently of the other similar states (Figure 2.11).

2.3.2 Transitions

A transition is an object that interconnects two states (Figure 2.5). The transition is drawn by dragging the mouse pointer, with the left button held down, from the point situated at a state image's border to the point situated at the alternative state image's border. To move any transition end, the following actions are made: (a) the mouse pointer is put onto the transition end to be dragged (so that the mouse pointer looks like a circle), and (b) the transition end is moved to the required position.

To change the transition branch shape, any of its point can be dragged by the mouse.

A transition has a label that describes the conditions of transition operation and the accompanying actions. Labeling is performed by double-clicking close to the transition branch. To move the transition label, it should be selected, and its new fixed position should be set by moving the mouse pointer.

The label text is formatted as follows:

```
event [condition]{condition_action}/transition_action,
```

where

1. **event:** Determines the transition-initiating event. If the event is not specified, the transition is initiated by performing the Boolean expression; if the transition-initiating events are numerous, all of them are specified and separated by an OR operator.

2. **condition:** The Boolean expression that initiates the transition when it becomes true. It also initiates certain actions (for instance, Figure 2.7 shows the transition with ascribed input operation condition = 0).

3. **condition_action:** The action executed after the transition condition becomes true but before the entire transition is defined as true (and the target state is determined).

4. **transition_action:** Determines the actions executed during transition in case all described transitions are true and the target state is already determined (for instance, for the transition in Figure 2.7, the generation of output signal + 0 is this kind of action).

2.3.3   Connective Junctions

A connective junction is a graphic object enabling simplification of Stateflow diagrams and generation of a more efficient code. A connective junction is a *no-happening* state: only the execution of the transition conditions is expected. This junction is employed when the state has more than one transition expecting the same event but operating with the true value of various Boolean expressions.

To draw a *branching* or *unification* type of connective junction in a Stateflow diagram, the following operations are performed: (a) select the Connective Junction icon on the toolbar of the Stateflow window, and (b) by clicking the left mouse button, indicate the connective junction's required position on the Stateflow diagram. The use of connective junctions in the case of a multivariate response to events is illustrated by Figure 2.12.

2.3.4   Default Transitions

A default transition is a transition with a preassigned target but without a preassigned source. It can be used as the starting state analog. It is also used

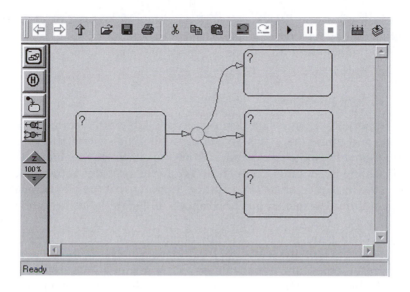

FIGURE 2.12A
Branching type of connective transition.

in hyperstate to determine which of the hyperstate-constituting states becomes active when the system enters this hyperstate. To set a default transition in a Stateflow diagram, the Default Transition icon should be selected on the toolbar of the Stateflow window, the mouse cursor placed on the outline of a future starting state, and the left mouse button clicked.

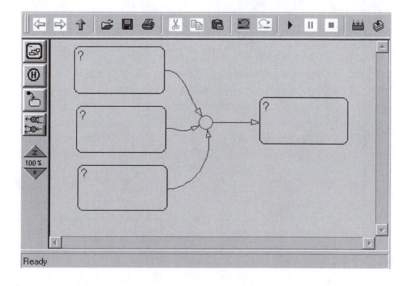

FIGURE 2.12B
Unification type of connective transition.

The default transition drawing was given above in Figure 2.6. The default transition editing (change of shape, placement, and target state) is similar to the editing of standard transitions (see Section 2.3.2).

2.3.5   History Junctions

Let us assume that the system leaves a hyperstate with a complex internal structure and numerous states. In a short while, the system returns and by definition must resume its operation from the starting state; the user, however, is willing to continue the operation from the previous active state. That is when the history junction is necessary. It operates at the hierarchical level of its definition. The history junction cancels all default transitions of a given hyperstate.

To set a history junction in a Stateflow diagram, the Default Transition icon should be selected on the toolbar of the Stateflow window, the mouse cursor placed on the box of the hyperstate in which it should be located, and the left mouse button clicked. There is no need to connect the element with any state by transition (Figure 2.13).

Double-clicking the left mouse button on the Default Transition icon assists in setting several history junctions without the need for selecting the corresponding icons. To choose cancel, clicking the right mouse right button is sufficient.

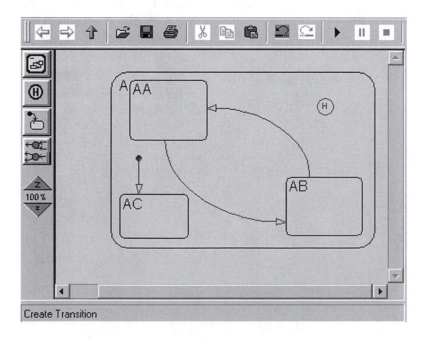

FIGURE 2.13
Generation of a history junction.

2.4 Defining Events and Data

Events are the Stateflow diagram's nongraphic objects, and they control the diagram. All events relating to a given Stateflow diagram should be defined. The types of events are input events, output events, local events, events imported from an external source, and events exported to the external target.

To create an event, the corresponding item should be selected in the menu of the Stateflow window:

1. **Add/Event/Input from Simulink:** Input events
2. **Add/Event/Output to Simulink:** Output events
3. **Add/Event/Local:** Local events

The event name and its alternative characteristics should be entered to the opened Event dialog window (Figure 2.14).

Variables are the Stateflow diagram nongraphic objects and are intended for numerical data storage. Variables can be employed at any hierarchical level. The following types of variables are used:

1. Input variables
2. Output variables
3. Local variables
4. Constants
5. Variables that exist within a certain time interval only

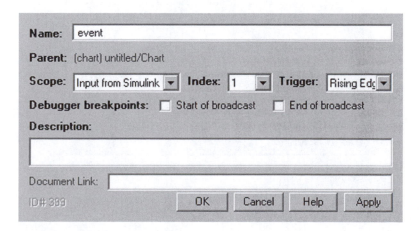

FIGURE 2.14
Generation of an input event for a Stateflow diagram.

6. Variables defined in the MATLAB® workspace
7. Variables imported to the target, which is external in Stateflow and Simulink

To create variables, the corresponding item should be selected in the menu of the Stateflow window:

1. **Add/Data/Input from Simulink:** Input variables
2. **Add/Data/Output to Simulink:** Output variables
3. **Add/Data/Local:** Local variables
4. **Add/Data/Constant:** Constants
5. **Add/Data/Temporary:** Existing within a certain time interval only

The variable name and its alternative characteristics should be entered to the opened Data dialog window (Figure 2.15).

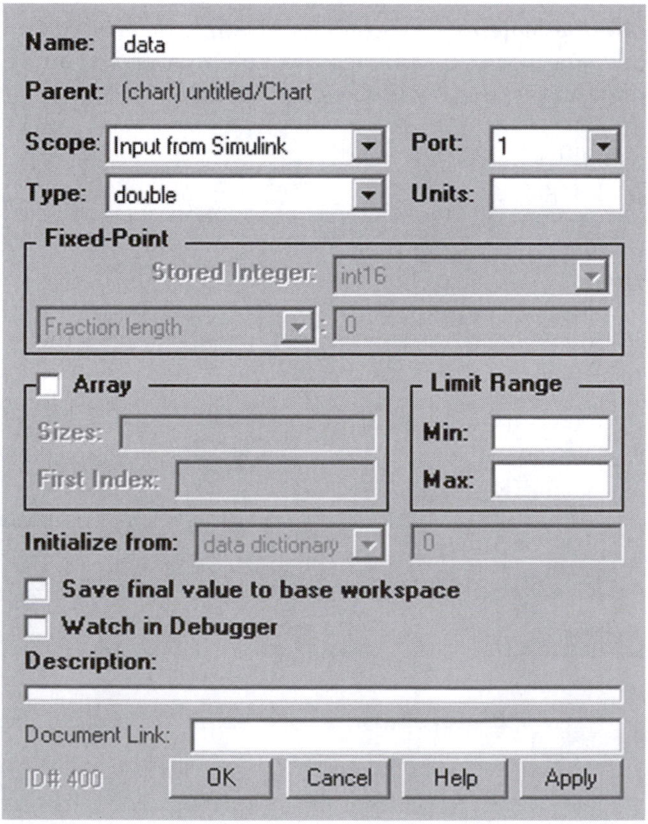

FIGURE 2.15
Generation of input variable for a Stateflow diagram.

2.5 Defining Stateflow Interfaces

2.5.1 Inputs

The two types of inputs from Simulink to Stateflow diagram are the data and the events. The first input type is the data input port. Each such port is intended for input data obtaining a Stateflow diagram input variable from the Simulink model. These input data can be transferred as a vector or a scalar. To set an input, the Data instruction is selected from the Add menu of the Stateflow window (or from the Stateflow Explorer window), whereas the second-level menu shows the Input from Simulink instruction (Figure 2.15).

All Stateflow diagram input events pass through the only input port. If several events are available, they should be combined into a vector signal using a multiplexor. The sequence of elements in a vector input signal corresponds to the sequence of Stateflow-assigned events.

To set an event, the Event instruction is selected from the Add menu of the Stateflow window (or from the Stateflow Explorer window), whereas in the second-level menu the Input from Simulink instruction is used (Figure 2.14). The Stateflow Explorer window permits the user to edit the sequence of input events by changing the event index in the corresponding dialog window.

2.5.2 Outputs

Stateflow diagrams have two types of outputs: data and events. The output data are set using the Output to Simulink instruction of the second-level menu (Figure 2.16), that can be accessed by the Data instruction from the Add menu of the Stateflow window (or the Stateflow Explorer window). The output data can also be transferred as a vector or a scalar. Each output variable is connected to a separate output, whose name corresponds to that of an output variable.

The output events are assigned using the Output to Simulink command of the second-level menu (Figure 2.17) that can be accessed by the Event instruction from the Add menu of the Stateflow window (or the Stateflow Explorer window).

2.6 Exploring and Searching

The Stateflow Explorer window permits the user to view all data and events in the same window, edit their properties, and declare new data and events. To open a Stateflow Explorer window, the Explore instruction is selected

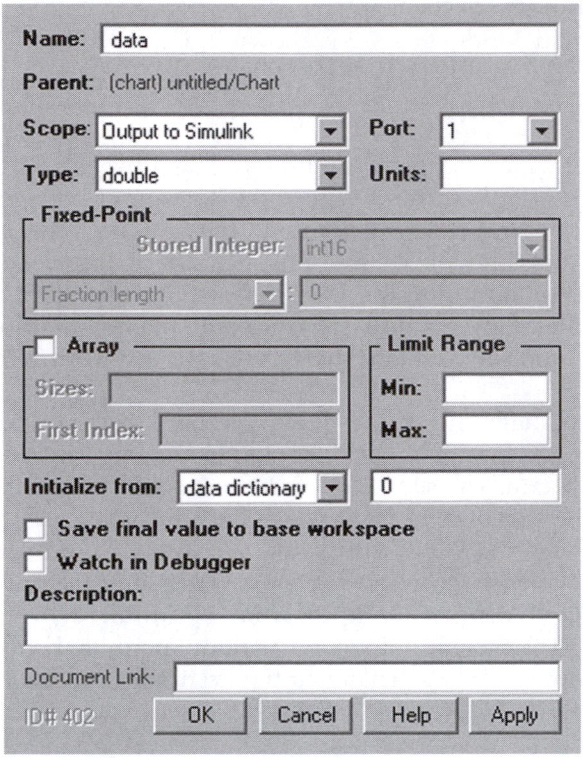

FIGURE 2.16
Generation of an output variable for a Stateflow diagram.

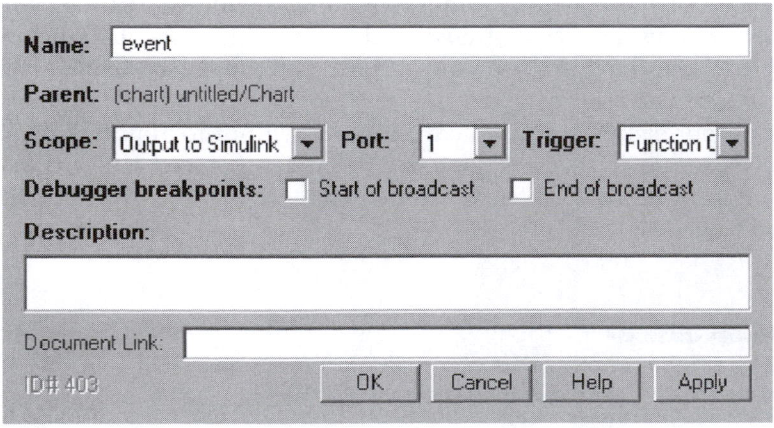

FIGURE 2.17
Generation of an output event for a Stateflow diagram.

from the Tools menu of the Stateflow window. The Stateflow Explorer window for the model in Section 2.2 is shown in Figure 2.8. It has two panels:

1. **Object Hierarchy:** Permits viewing of the hierarchy of all Stateflow diagrams currently opened.
2. **Contents of:** Contains a list of data and events assigned for the Stateflow diagram selected in the Object Hierarchy panel.

To edit the properties of the data or event element, the left mouse button should be double-clicked on the required element in the list and the Properties instruction selected from the Edit menu of the Stateflow Explorer window. S dialog window of the element properties will be displayed on the screen.

For assigning a new variable (data element) or a new event, a respective instruction from the Add menu of the Stateflow Explorer window should be selected (its content being identical to the Add menu of the Stateflow window).

To search the Stateflow diagram objects, a navigator is used, which is called by an instruction from the Tools/Find menu of the Stateflow window. This instruction opens the object search window, wherein the name of the object to be found can be given. Pressing the Find button displays the search results: the list of all objects with a given name appears at the bottom of the search window.

2.7 Debugging

The Stateflow software has developed the means of program debugging, and the most vital is the Stateflow Builder syntax analyzer that can be started by the instruction from the Tools/Parse menu in the Stateflow window. The Stateflow/Parser window contains data on current the Stateflow diagram and on the presence of syntax errors, if any (and their nature and sources). When the syntax control is over, a dialog window with diagnostic notifications and error sources appears. If syntax errors are absent, the diagnostic notification shows only one word: Done.

Absence of syntax errors cannot ensure the fault-free operation of Stateflow diagram because the diagram can contain semantic errors, which are more serious. Stateflow has the debugging means for semantic error detection, and offers the possibility of step-by-step tracing of the Stateflow diagram execution logic; animation can be used to upgrade the operation pictorial information.

At the debugging stage, a stand-alone Stateflow diagram startup can be performed, following the Start instruction from the Simulation menu of the Stateflow window. It is expedient to check the model parameter values

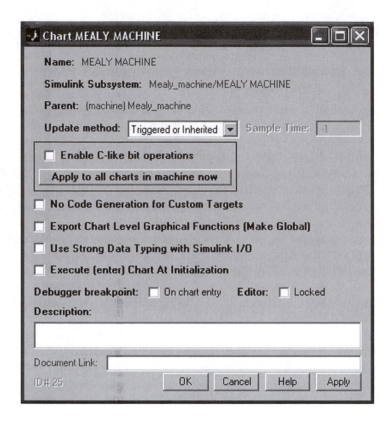

FIGURE 2.18
The Dialog window for setting the model parameters.

before the instruction execution. The values are set using a dialog window (Figure 2.18) that opens by selecting the Chart Properties instruction in the Stateflow window File menu.

The window contains the following elements:

1. **Name**: The Stateflow diagram name. This window is a static textbox implemented as a hyperlink; mouse clicking on the textbox activates the Stateflow diagram window.

2. **Simulink Subsystem:** The Simulink model name that also includes the Stateflow diagram. This is implemented as a hyperlink as well.

3. **Parent:** The higher-level Stateflow diagram name (parent). This is implemented as a hyperlink.

4. **Update method**: This opening list permits the user to select the control technique for Stateflow diagram operation.

   Triggered or Inherited: This technique is used when the Stateflow diagram is controlled by the events or the data incoming from a

Simulink model: if the Stateflow diagram input events are defined, it will be activated each time a controlling signal arrives at the Stateflow diagram triggered input; if only the input data are defined, the Stateflow diagram activation periodicity is dictated by the data input intensity; if any information or event links with Simulink are absent, the diagram periodicity is equal to the simulation step set for a Simulink model.

Sampled: The Stateflow diagram activation periodicity is assigned as the Sample Time parameter value for the respective block; in this process Simulink generates the controlling events at equal time intervals corresponding to the parameter value.

Continuous: The Stateflow diagram is activated at each simulation step, its value being defined by the Simulink model parameters.

5. **Use Strong Data Typing with Simulink I/O:** This checkbox allows banning or permission of utilization of data stored by Simulink in the MATLAB workspace.

6. **Execute (enter) Chart At Initialization:** This checkbox allows banning or permission of Stateflow diagram execution following the Simulink model start.

7. **Debugger breakpoint:** This checkbox allows using the Stateflow diagram initialization point as a halt point during debugging.

8. **Editor:** This checkbox allows banning any Stateflow diagram modification.

For the Stateflow diagram stand-alone startup, the following operations should be made:

1. Open the Stateflow diagram window.
2. Select the Open Simulation Target instruction in the Tools menu of the Stateflow window.
3. When the Simulation Target Builder dialog window opens (Figure 2.19), click the mouse on the Coder Options button and check the Enable Debugging/Animation checkbox in the add-in window (Figure 2.20).
4. Choose the Debug instruction in the Stateflow window Tools menu and check the Enable switch in the Animation group in the Stateflow Debugging window (Figure 2.21).
5. Click the Start button in the Stateflow Debugging window.

In conclusion, we give the example of FSM model building in the Stateflow environment, using ASM (algorithmic state machine) as the alternative FSM representation. The starting data is the block diagram of the machine operation algorithm, the ASM chart (Figure 2.22). The simulation of ASM faults is dealt with in Chapter 3.

FIGURE 2.19
The Simulation Target Builder window.

The ASM chart block diagram [6–8] is constructed by means of the following standard boxes: the system state box (Figure 2.23), the decision box (Figure 2.24), and the conditional output box (Figure 2.25).

All the machine states are implemented in the definite number of system state blocks. The machine output signals are subdivided into two types [7]: (a) Moore outputs, which are connected to the system state blocks, for

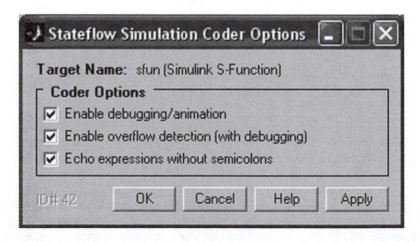

FIGURE 2.20
The Stateflow Simulation Coder Options window.

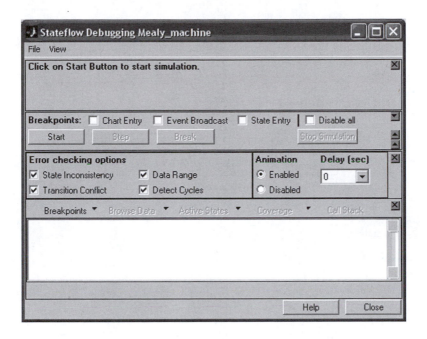

FIGURE 2.21
The Stateflow Debugging window.

instance, .ADD in Figure 2.23, so that means given signal is active in the logical unity state) if the signal is written as L.ADD, it means the signal is active in logical zero state), and (b) Mealy outputs, which are connected to the state-to-state transitions, for instance, IL.CLOCK, IH.RESET in Figure 2.25; the letter I means the given output signal is a Mealy output).

The machine starting state is denoted by any of the three possible methods shown in Figure 2.26 [7]. The machine output signals are analyzed using the

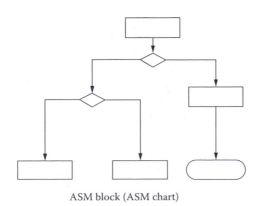

ASM block (ASM chart)

FIGURE 2.22
The ASM chart.

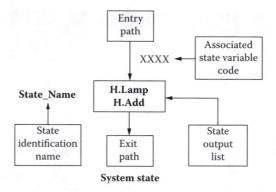

System state

FIGURE 2.23
The system state box.

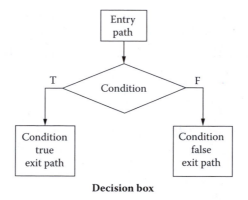

Decision box

FIGURE 2.24
The decision box.

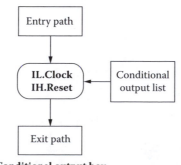

Conditional output box

FIGURE 2.25
The conditional output box.

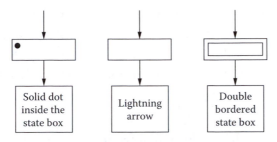

Reset or starting states

| Solid dot inside the state box | Lightning arrow | Double bordered state box |

FIGURE 2.26
Reset or starting states.

decision boxes. For instance, the ASM chart block diagram in Figure 2.27 [7] contains four states A, B, C, and D, and eight possible paths of interstate transitions:

1. **Path L1:** Transition from state A to state B. It is implemented at the following input signal combination: 1 = '0', X3 = '0'.
2. **Path L2:** Transition from state A to state A, provided that 1 = '1', X2 = '0'.
3. **Path L3:** Transition from state A to state C, provided that 1 = '0', X2 = '0', X3 = '1'.
4. **Path L4:** Transition from state A to state C, provided that 1 = '0', X3 = '1'.

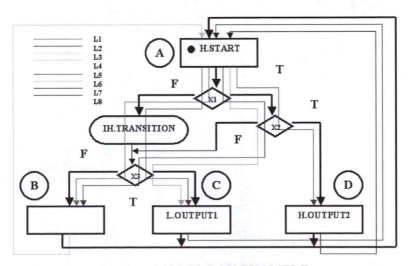

ASM BLOCK EXAMPLE

FIGURE 2.27
An ASM Block example.

5. **Path L5:** Transition from state A to state A, provided that 1 = '1', X2 = '1'.

6. **Path L6:** Unconditional transition from state B to state A.

7. **Path L7:** Unconditional transition from state C to state A.

8. **Path L8:** Unconditional transition from state D to state A.

The machine's state A is its starting state (marked in Figure 2.26 with the aid of way 1). The machine has three input signals (1, 2, 3) and four output signals. Three of them are Moore outputs related to the following machine states: (a) H.START is related to state A; (b) L.OUTPUT1 is related to state C;

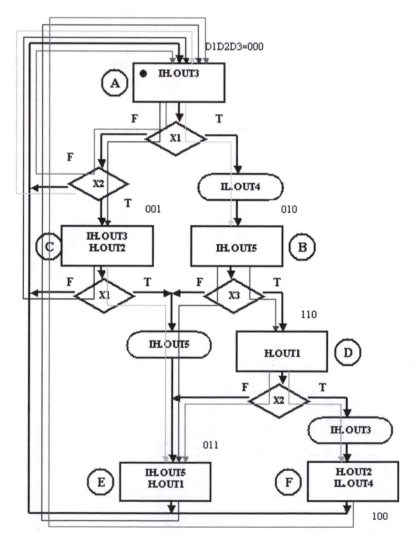

FIGURE 2.28
An ASM chart.

(c) .OUTPUT2 is related to state D. The fourth output signal is the Mealy output related to the transition from state A to state B or C (IH.TRANSITION). Note that the Mealy output can also be recorded inside any of the system state rectangles. It will signify that this Mealy output is connected to all state-outgoing paths.

Figure 2.28 shows the diagram of ASM chart states, transitions, input, and output signals that contains [7]:

1. Six states (A, B, C, D, E, F)
2. Three input signals (X1, X2, X3)
3. Eleven interstate transition paths (linkpaths) (L1...L11)
4. Five output signals, where two are the Moore outputs (H.OUT1, H.OUT2) and the remaining three are the Mealy outputs (IH.OUT3, IL.OUT4, IH.OUT5)

Based on this block diagram, a table called the Symbolic ASM Table can be plotted (Figure 2.29). Each table row corresponds to a path (linkpath). Using the state and the signal codes, one more table can be constructed: the Assigned ASM Table (Figure 2.30). The Stateflow diagram (Figure 2.31) is plotted on the basis of the Symbolic ASM Table or the Assigned ASM Table.

Symbolic ASM table

Link path	Inputs			Present state	Next state	Moore outputs		Mealy outputs		
	X1	X2	X3			H.out1	H.out2	IH.out3	IL.out4	IH.out5
L1	F	F	-	A	A	-	-	Active	-	-
L2	T	-	-	A	B	-	-	Active	Active	-
L3	F	T	-	A	C	-	-	Active	-	-
L4	-	-	F	B	D	-	-	-	-	-
L5	-	-	T	B	E	-	-	-	-	Active
L6	F	-	-	C	A	-	Active	Active	-	-
L7	T	-	-	C	E	-	Active	Active	-	Active
L8	-	F	-	D	E	Active	-	-	-	-
L9	-	T	-	D	F	Active	-	Active	-	-
L10	-	-	-	E	A	Active	-	-	-	Active
L11	-	-	-	F	A	-	Active	-	Active	-

F-False
T-True

FIGURE 2.29
The Symbolic ASM Table.

Assigned ASM table

Link-path	Inputs			Present state	Next state	Moore outputs		Mealy outputs		
	X1	X2	X3	D1 D2 D3	D1+D2+D3	H.Out1	H.Out2	IH.Out3	IL.Out4	IH.Out5
L1	0	0	-	0 0 0	0 0 0	0	0	1	1	0
L2	1	-	-	0 0 0	0 1 0	0	0	1	0	0
L3	0	1	-	0 0 0	0 0 1	0	0	1	1	0
L4	-	-	0	0 1 0	1 1 0	0	0	0	1	0
L5	-	-	1	0 1 0	0 1 1	0	0	0	1	1
L6	0	-	-	0 0 1	0 0 0	0	1	1	1	0
L7	1	-	-	0 0 1	0 1 1	0	1	1	1	1
L8	-	0	-	1 1 0	0 1 1	1	0	0	1	0
L9	-	1	-	1 1 0	1 0 0	1	0	1	1	0
L10	-	-	-	0 1 1	0 0 0	1	0	0	1	1
L11	-	-	-	1 0 0	0 0 0	0	1	0	0	0

FIGURE 2.30
The Assigned ASM Table.

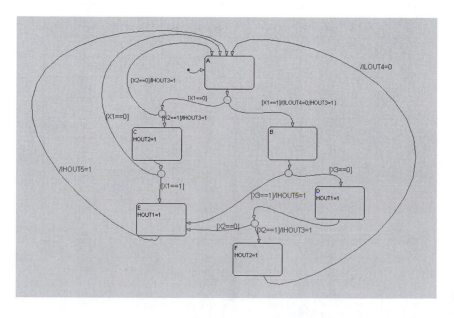

FIGURE 2.31
The Stateflow diagram for the ASM chart in Figure 2.28.

References

1. Harel D. Statecharts: a visual formalism for complex systems. *Science of Computer Programming*, 8(3), 231–274, 1987.
2. Harel D. On visual formalisms. *Communications of the ACM*, 31(5), 514–530, 1988.
3. *Stateflow User's Guide*. The MathWorks, Inc. Natick, MA, 2004.
4. Rumbaugh J., Jacobson I., Booch G. *The Unified Modeling Language Reference Manual*. Addison-Wesley-Longman, Reading, MA, 2001.
5. Dabney, J.B., Harman, T.L. *Mastering Simulink 4* (2nd ed.). Prentice Hall, New York, 2001.
6. Lee S. *Design of Computers and Other Complex Digital Devices*. Prentice Hall, New York, 1999.
7. Pool N.R. Lectures for Course 312EE Advanced Digital Systems (Algorithmic State Machines). Coventry University, U.K., 1999.
8. Perelroyzen E.Z. *VHDL Design* (in Russian). Solon-Press, Moscow, 2004.

3

Fault Modeling and Simulation

3.1 Fault Modeling

3.1.1 Fault Models for Combinational Circuits

Let us view the two-input logical NOR gate that implements the Boolean function:

$$f(x_1, x_2) = not(x_1 + x_2).$$

Owing to some internal electric fault, the gate, instead of this function can implement only one of the following:

$$f_1(x_1, x_2) = 0,$$

$$f_2(x_1, x_2) = x_1,$$

$$f_3(x_1, x_2) = x_2,$$

$$f_4(x_1, x_2) = 1.$$

These possibilities are given by the $F(f)$ fault model in the NOR gate:

$$F(f) = \{f_0 = f, f_1, f_2, f_3, f_4\},$$

where $f_0 = f$ is the gate normal function, and $f_1, f_2, f_3,$ and f_4 are the fault functions for the $F(f)$ fault model in the gate.

If any fault function of the fault model is implemented by the gate, it means that the gate has an engineering fault.

In practice, constant faults of the s@0 (stuck-at 0) and s@1 (stuck-at 1) type are of particular importance (stuck-at fault model). With the s@0 type of fault at the gate (or flip-flop) input or output, the signal of the given circuit node has the constant value of logical zero, independent of the input impacts

TABLE 3.1

Example of Fault Statistics for a Gate

f_i	0	x_1	x_2	1	Alternative functions
$P(f_i)$	50%	15%	15%	10%	10%

applied. Such constant faults are caused, for instance, by the grounding of a corresponding conductor. The s@0 type of fault by the output results in the realization of the $f_1(x_1, x_2) = 0$ function by the faulty gate. The s@0 type of fault by the input results in the realization of the $f_3(x_1, x_2) = x_2$ or $f_2(x_1, x_2) = x_1$ function by the faulty gate. The s@1 type of fault by the output results in the realization of the $f_4(x_1, x_2) = 1$ function by the faulty gate. The s@1 type of fault by the input results in the realization of the $f_1(x_1, x_2) = 0$ function by the faulty gate.

Engineering faults can occur with varying frequencies defined by the fault statistics. Each f_i function of the fault model has a corresponding $P(f_i)$ number expressed as a percent. $P(f_i)$ is the percentage of faults that results in $f_i \in F(f) \setminus \{f_0\}$ function implementation by the defective gate.

The example of fault statistics for such a gate is given in Table 3.1. It can be seen from the table that the fault model for our gate type covers 90% of all possible faults. On the basis of fault models for gates and flip-flops, the fault models for combinational circuits and sequential circuits can be obtained [1].

3.1.2   Fault Models for Sequential Circuits

A finite state machine (FSM) is known to be the sequential circuit model. The fully defined and determined initial Mealy machine, $A = (X, Y, Z, z_0, \delta, \lambda)$, is defined by:

1. The input and output sets X, Y
2. The set of states Z and initial state z_0
3. The transition functions δ: $Z \times X \rightarrow Z$
4. The output function λ: $Z \times X \rightarrow Y$

that are known for all $\in$ Z, x $\in$ X.

The A machine operates step-by-step and in synchronism with regard to the discrete periods of time t(t = 0, 1, 2,...). For z(t), z(t + 1) $\in$ Z, x(t) $\in$ X, y(t) $\in$ Y the following relations are held:

$$z(t + 1) = \delta(z(t), x(t));$$

$$y(t) = \lambda(z(t), x(t)).$$

We denote the set of all finite words out of X (Y) set by X^* (Y^*) so that Λ is an empty word that does not contain a single element.

The augmented transitions function,

$$\delta: X \times X^* \to Z,$$

and augmented output function,

$$\lambda: Z \times X^* \to Y,$$

for $p \in X^*$, $x \in X$, and $z \in Z$ are defined from induction,

$$\delta(z, \Lambda) = z; \quad \lambda(z, \Lambda) = \Lambda;$$

$$\delta(z, px) = \delta(\delta(z, p), x); \quad \lambda(z, px) = \lambda(\delta(z, p), x),$$

and are denoted by the same letters as the initial transitions and outputs functions.

Transformation of input words into output words made by the A machine is described by the λ^* function. For $x_1, x_2, \ldots, x_n \in X^*$ and $y_1, y_2, \ldots, y_n \in Y^*$, the equality $\lambda^*(z, x_1, x_2, \ldots, x_n) = \lambda(z, x_1) \lambda(\delta(z, x_1), x_2) \ldots \lambda(\delta(z, x_1), x_2) \ldots \lambda(z, x_1, x_2, \ldots, x_{n-1}), x_n) = y_1, y_2, \ldots, y_n$ is held.

The $A' = (X', Y', Z', \delta', \lambda')$ machine is called the submachine of the $A = (X, Y, Z, \delta, \lambda)$ machine if

1. $X' \subseteq X, Y' \subseteq Y, Z' \subseteq Z.$
2. For all $x \in X', y \in Y', z \in Z'$ the relationships

$$\delta'(z, x) = \delta(z, x),$$

$$\lambda'(z, x) = \lambda(z, x)$$

are held (i.e., the δ and λ functions on the $Z'^* X'$ set coincide the with δ' and λ' functions).

The states with similar input and output values are called the equivalent states. If $A = (X, Y, Z, \delta, \lambda)$ and $L = (X, Y, Z', \delta', \lambda')$ are machines, then $z \in Z$ and $z' \in Z'$ are equivalent if the $\lambda(z, p) = \lambda'(z', x)$ relationship is held for all $p \in X^*$.

The A and L machines are equivalent if $z' \in Z'$ can be found for each $z \in Z$, and if $z \in Z$ can be found for each $z' \in Z'$ so that z and z' are equivalent. For $A = L$, two $z, z' \in Z$ states are equivalent if for all $p \in X^*$ the $\lambda(z, p) = \lambda(z', p)$ relationship holds.

The equivalence of states defines certain partitions by the equivalence ratio on the set of automaton states. The equivalent states are united into the equivalence classes. The machines without a single pair of equivalent states are called reduced machines (automata with a minimal number of states). In such machines, each class of the equivalent states comprises only one state. Having replaced all states in each equivalence class by a single state, we produce the reduced machine out of the equivalent nonreduced one.

In practice, the machines are frequently only partially defined. For many input sets and states, the analysis of machine behavior is meaningless, since the corresponding *input set-state* combinations do not take place in practice, or alternately, the machine behavior is of no interest when a certain state has been achieved (though for any real sequential circuit the machine operates in a definite manner even at these undefined combinations).

If $A = (X, Y, Z, \delta, \lambda)$ and the subsequent state z for $z \in Z$, $x \in X$ is not defined, then $\delta(z, x) = $ the underdefined state at the logical level, and with undefined output, $\lambda(z, x) = $ the undefined state.

If a partially defined A machine is implemented as the sequential circuit, this sequential circuit behavior coincides with the behavior of a partially defined machine for the input set-state combinations that served as the basis for the machine definition. For the input set-state combinations with the undefined machine operations, the sequential circuit operation is set arbitrarily. When a sequential circuit is implemented with minimal costs, the undefined values are redefined so that the sequential circuit has a minimal number of states.

Equivalent states and the classes of equivalent states in fully defined machines correspond to the compatible states and to the maximal classes of compatible states in partially defined machines. If $A = (X, Y, Z, \delta, \lambda)$ and $A = (X, Y, Z, \delta', \lambda')$ are partially defined machines, then the $z \in Z(A)$ and $z' \in Z'(L)$ states are compatible if for all $p \in X^*$, whose input data are defined both for z and z', the following condition is met:

$$\lambda (z, p) = \lambda (z', p),$$

on the condition that both output functions are defined for z and z'.

For $A = L$, both A states are called compatible if for all x $p \in X'$ defined for these states, the following equity is obeyed:

$$\lambda (z, p) = \lambda' (z', p),$$

on the condition that both λ, λ' are defined for z, z'.

We will examine the sequential circuit that implements the four-state Gray counter (Figure 3.1) [1]. We proceed from the following assumptions:

1. The s@0 type of fault at the element output is taken to be the only cause of faults for all elements (gates and flip-flops).

2. At each moment of time one element at most operates in a faulty way. This assumption is rational, since the mean time between failures for an individual element is up to 10^{10} hours; hence, the occurrence of a second error prior to detection of the first error is hardly likely.

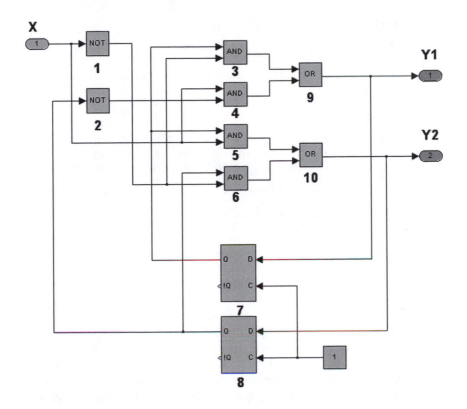

FIGURE 3.1
Sequential circuit that implements the four-state Gray counter.

Our sequential circuit elements possess the following fault models:

1. Inverter: $F(\bar{x}) = \{\bar{x}, 0\}$
2. AND gate: $F(x_1, x_2) = \{x_1 x_2, 0\}$
3. OR gate: $F(x_1 + x_2) = \{x_1 + x_2, 0\}$
4. D flip-flop: $F(x(t-1)) = \{x(t-1), 0\}$

For an $A_G = (X = \{0, 1\}$ machine,

$$Z = Y = \{00, 01, 10, 11\},$$

$$z_0 = 00,$$

$$\delta = (\delta_1, \delta_2),$$

$$\lambda = \delta$$

TABLE 3.2

δ_i Functions of A_i ($i = 1, 2,..., 10$) Machines

x	z	δ	δ_1	δ_2	δ_3	δ_4	δ_5	δ_6	δ_7	δ_8	δ_9	δ_{10}
0	00	00	00	00	00	00	00	00	00	00	00	00
0	01	01	00	01	01	01	01	00	01	00	01	00
0	10	10	00	10	00	10	10	10	00	10	00	10
0	11	11	00	11	01	11	11	10	01	11	01	10
1	00	01	01	00	01	01	00	01	01	00	01	00
1	01	11	11	10	11	01	10	11	01	10	01	10
1	10	00	00	00	00	00	00	00	00	00	00	00
1	11	10	10	10	10	00	10	10	11	10	00	10

the following conditions are met:

$$y_1(t) = z_1(t + 1) = x(t)\,\bar{z}_2(t) + \bar{x}(t)z_1(t);$$

$$y_2(t) = z_2(t + 1) = x(t)\,z_1(t) + \bar{x}(t)z_1(t).$$

The A_i machines of the $F(A_G)$ fault model are produced from A_G by successive replacement of the starting circuit elements (Figure 3.1) for the modified elements and performance analysis of the respective modified sequential circuits. For example, we define the A_1 machine for the fault model. Here, the inverter (the sequential circuit element 1) should replace the element that is the constant logical zero source. Having analyzed the modified sequential circuit for A_1, we have:

$$y_1(t) = z_1(t + 1) = x(t)\,\bar{z}_2(t) = \delta_{11}(x(t),\, z_1(t),\, z_2(t));$$

$$y_2(t) = z_2(t + 1) = x(t)\,z_1(t) = \delta_{12}(x(t),\, z_1(t),\, z_2(t)).$$

Hence, the values of $\delta_1 = (\delta_{11},\, \delta_{12}) = \lambda_1 = (\lambda_{11},\, \lambda_{12})$ can be obtained.

By successive replacement of the sequential circuit elements (second to tenth) for the modified elements, the sources of permanent logical zero, we obtain the modified sequential circuit machines from A_2 to A_{10} of the $F(A_G)$ fault model for the Gray counter. The δ_i functions of A_i ($i = 1, 2,..., 10$) machines are listed in Table 3.2.

For feasible large-size combinational circuits and sequential circuits, the fault models are defined using the fault simulation system.

3.2   Fault Simulation

3.2.1   Fault Simulation for Combinational Circuits

We will now examine the troubleshooting technique for any single fault (the Sogomonyan technique) that allows for the introduction of the F fault model for the f_s combinational circuit that implements the f: $\{0,1\}^n \rightarrow \{0,1\}^m$; n, m ≥ 0 function in such a manner that Φ_s troubleshooting circuit detects all plausible single faults and the fault model contains a minimal amount of fault functions [1].

This technique stems from the fact that each ombinational circuit is implemented by connections of the G_i (i = 1, 2,...) gates. Since feedbacks are absent in combinational circuits, the following partial ordering is defined for the set of all combinational circuit Γ_G gates: the $G_i < G_j$ condition is strictly obeyed when the connection runs from G_i to G_j. Also, the $G_i \leq G_j$ condition holds true for all combinational circuit gates $G_i \in \Gamma_G$. On the set of Γ_G gates a class subdivision is defined when each class comprises its lower border and at the same time such subdivision cannot be amalgamated. Such a class subdivision is defined unambiguously.

The F fault model for combinational circuits contains the number of fault functions f_1, f_2,... identical to the number of C_G classes. The fault functions f_i are defined as follows: the class boundary gates are successively replaced for the inverted gates, and the functions implemented by the thus modified ombinational circuit are the fault functions of the fault model. Thus, each fault function of the fault model is applied to all faults of C_G class gates and encompasses all plausible single faults of the given class.

Plausible faults of the gates pertaining to the same C_G class should be detected at the output of the C_G class boundary gates if they can be in general detected at the combinational circuit output. If the combinational circuit has only one output, said technique results in common duplication and comparison.

The C_G gate classes and their boundary gates are defined as follows. The combinational circuit f_s gates are numbered from 1 to n. In the table relating gates and combinational circuit outputs to n lines, n + 1 columns, a unity is recorded in the (i, j) cell for the case when the connection between G_i and G_i (i $\neq$ j) is found in f_s combinational circuits. A unity is recorded in the (k, n + 1) cell only when the G_k gate is connected directly to the combinational circuit (1 $\leq$ i, j, k $\leq$ n) output.

The definition algorithm for C_G classes and their boundary gates is composed of the following steps:

1. A table row where a unity stands only in the (n + 1)th column and has the ** sign. (Such a row exists because f_s has no feedbacks.)

2. A column whose number coincides with the number of the already labeled row and has the * sign.

3. The row that contains the unities only in the labeled columns is found. If only one such row is found, it is marked with the * sign and the transition to step 4 is made; otherwise, step 2 is executed.

4. The labeled rows and columns are crossed out of the table (the gates with crossed-out numbers form the C_G class; the gate whose number is equal to the **-marked row number is the boundary gate).

5. If a unity pertaining to the unlabeled row is deleted by crossing out a column, then a unity is added in the $(n + 1)$th column of the same row.

6. If alternative undeleted rows are still found, the transition to step 1 is made.

The algorithm is over when all table rows are crossed out.

Example 1 [1]

Figure 3.2 gives the combinational circuit for a 1-bit adder.

The given combinational circuit gates are numbered 1 to 12. Therefore, the table of combinational circuit gates and output connections presents 12 rows and 13 columns (Table 3.3). The (i, j) cell contains 1 if the i element output is connected to the j element input. The $(i, 13)$ cell contains 1 if the i element is connected to the circuit output. Following this algorithm, we execute steps 1 to 4.

1. We choose row 11 as the first row, having a unity only in column 13, and mark it with **.

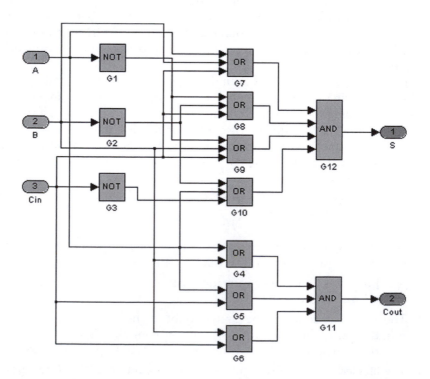

FIGURE 3.2
The combinational circuit for a 1-bit adder.

TABLE 3.3

Table of Connections for Example 1

	1	2	3	4*	5*	6*	7	8	9	10	11*	12	13
1								1	1				
2								1		1			
3									1	1			
4*											1		
5*											1		
6*											1		
7												1	
8												1	
9												1	
10												1	
11**													1
12													1

2. We mark column 11 with *. The column contains the unities in rows 4, 5, and 6.

3. Row 4 has a unity only in the labeled column 11. Therefore, we mark the row and column 4 with *. In the same manner we label rows and columns 5 and 6. Gates 4, 5, 6, and 11 form the $C_{G1} = \{4, 5, 6, \underline{11}\}$ class with boundary gate 11 (its number is underlined).

4. The labeled rows and columns are crossed out. Then we have the updated table of connections (Table 3.4).

We again apply the algorithm to the updated Table 3.4.

1. Only row 12 contains a unity in column 13. It is marked with **.

2. Label column 12 with *. The column contains unities in rows 7, 8, 9, and 10.

3. Row 7 contains a unity only in the labeled column 12. Hence, we label the row and the column 7 with *. Column 7 does not contain a unity, so we proceed to column 8 and label it, as well as row 8. Column 8 contains unities in rows 1 and 2. These rows contain additional unities in the so far unlabelled columns 9 and 10. Therefore, these rows do not satisfy the

TABLE 3.4

Updated Table of Connections for Example 1

	1*	2*	3*	7*	8*	9*	10*	12*	13
1*					1	1			
2*					1		1		
3*						1	1		
7*								1	
8*								1	
9*								1	
10*								1	
12*									1

step 3 conditions, and we proceed to column 9 and label it, as well as row 9. This column contains unities in rows 1 and 3. Row 1 contains unities in the labeled columns only, so both the row and column 1 are labeled *. Now we proceed to column 10 and label it, as well as row 10. Column 10 contains unities in rows 2 and 3, which now contain unities only in the labeled columns; therefore, both they and the corresponding columns 2 and 3 are labeled *.

4. All table rows are now labeled. It follows that the 1, 2, 3, 7, 8, 9, 10, and 12 gates combine into the $C_{G2} = \{1, 2, 3, 7, 8, 9, 10, \underline{12}\}$ class with border gate 12 (its number is underlined). All labeled rows and columns are crossed out, and then the connections table does not have either rows or columns. The class generation has terminated.

Now we will define the fault model for a given $F(f_s)$ combinational circuit. The fault model will contain three fault functions (one of them, f_{s0}, corresponds to the combinational circuit's regular performance, whereas two others, f_{s1}, f_{s2}, describe the performances of modified combinational circuits, where the border gates 11 and 12 of C_{G1} and C_{G2} classes have been replaced by the inverter gates (here we deal with the output sum inversion and the 1-bit adder carry):

$$f_{sj} = (S'_j, C' \text{ out}_j); j = 0, 1, 2;$$

where

$$S'_0 = S_0, \qquad C' \text{ out}_0 = Cout_0;$$

$$S'_1 = S_0, \qquad C' \text{ out}_1 = \overline{Cout_0};$$

$$S'_2 = \overline{S_0}, \qquad C' \text{ out}_2 = Cout_0;$$

and

$$F(f_s) = \{f_{s0}, f_{s1}, f_{s2}\}.$$

The $F(f_s)$-parameters are given in Table 3.5.

TABLE 3.5

$F(f_s)$- Parameters for Example 1

A	B	Cin	S'_0	C'out$_0$	S'_1	C'out$_1$	S'_2	C'out$_2$
0	0	0	0	0	0	1	1	0
0	0	1	1	0	1	1	0	0
0	1	0	1	0	1	1	0	0
0	1	1	0	1	0	0	1	1
1	0	0	1	0	1	1	0	0
1	0	1	0	1	0	0	1	1
1	1	0	0	1	0	0	1	1
1	1	1	1	1	1	0	0	1

TABLE 3.6

Values for the Fault Detection Functions Φ_{si} for Example 1

A	B	Cin	S	Cout	Φ_{s1}	Φ_{s2}
0	0	0	0	0	0	0
0	0	1	1	0	0	0
0	1	0	1	0	0	0
0	1	1	0	1	0	0
1	0	0	1	0	0	0
1	0	1	0	1	0	0
1	1	0	0	1	0	0
1	1	1	1	1	0	0
0	0	0	0	1	1	1
0	0	1	1	1	1	1
0	1	0	1	1	1	1
0	1	1	0	0	1	1
1	0	0	1	1	1	1
1	0	1	0	0	1	1
1	1	0	0	0	1	1
1	1	1	1	0	1	1
0	0	0	1	0	1	1
0	0	1	0	0	1	1
0	1	0	0	0	1	1
0	1	1	1	1	1	1
1	0	0	0	0	1	1
1	0	1	1	1	1	1
1	1	0	1	1	1	1
1	1	1	0	1	1	1
0	0	0	1	1	–	–
0	0	1	0	1	–	–
0	1	0	0	1	–	–
0	1	1	1	0	–	–
1	0	0	0	1	–	–
1	0	1	1	0	–	–
1	1	0	1	0	–	–
1	1	1	0	0	–	–

From here, we can plot the table of values for the fault detection functions Φ_{si} (they are defined from the single-fault condition at the circuit outputs) (Table 3.6).

Example 2 [1]

In this example the described technique results in duplication and comparison. Again, we have the 1-bit adder, which is only alternately implemented (Figure 3.3).

In conformity with the algorithm described above we obtain the following gate classes with the underlined boundary gates (Table 3.7): $C_{G1} = \{8, \underline{11}\}$, $C_{G2} = \{1, 6, 7, 9, 10, 12, 13, \underline{14}\}$, $C_{G3} = \{\underline{4}\}$, $C_{G4} = \{\underline{2}\}$, $C_{G5} = \{\underline{3}\}$, and $C_{G6} = \{\underline{5}\}$.

The fault model $F(f_s)$, in addition to the function describing the operable combinational circuit performance,

$$f_{s0} = (S0, Cout0)$$

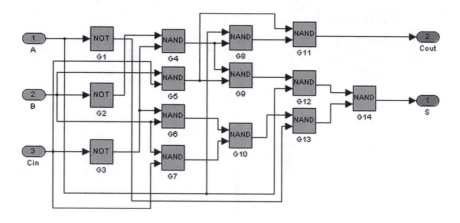

FIGURE 3.3
Alternative realization of the 1-bit adder.

includes six fault functions:

$$f_{si} = (Si, Couti); i = 1,\ldots, 6.$$

The fault functions f_{si} are obtained from the initial combinational circuit by inversion of the class border gates (the NAND gates 11, 14, 4, 2, 3, 5 are substituted by AND gates). The $f_{s0}, f_{s1}, f_{s2}, f_{s3}, f_{s4}, f_{s5}$, and f_{s6} values are listed in Table 3.8. The table assists in defining the Φ_s fault detection function. In this case we have border gates that functionally affect both adder outputs. If these NAND gates 4, 2, 3, and 5 are replaced with AND gates, the faults can affect both outputs of a 1-bit adder.

TABLE 3.7

Table of Connections for Example 1

	1	2	3	4	5	6	7	8	9	10	11	12	13	14	15
1													1		
2				1			1								
3				1		1									
4								1	1						
5									1		1				
6										1					
7										1					
8											1				
9												1			
10													1		
11														1	
12													1		
13													1		
14															1

TABLE 3.8

F(f$_s$)-Parameters for Example 2

A	B	Cin	S$_0$	C$_0$	S$_1$	C$_1$	S$_2$	C$_2$	S$_3$	C$_3$	S$_4$	C$_4$	S$_5$	C$_5$	S$_6$	C$_6$
0	0	0	0	0	0	1	1	0	0	0	0	0	0	0	0	1
0	0	1	1	0	1	1	0	0	1	0	0	0	1	0	1	1
0	1	0	1	0	1	1	0	0	1	0	1	0	0	0	1	1
0	1	1	0	1	0	0	1	1	0	1	1	1	1	1	0	0
1	0	0	1	0	1	1	0	0	0	1	0	1	0	1	1	1
1	0	1	0	1	0	0	1	1	1	0	0	1	1	0	1	1
1	1	0	0	1	0	0	1	1	1	0	1	0	0	1	1	1
1	1	1	1	1	1	0	0	1	1	1	1	1	1	1	0	1

The fact that the Φ_s function has only five undefined values can be checked. Thus, no Φ_s can be implemented with lower expenditures than the duplication of a given combinational circuit for the 1-bit adder and subsequent comparison of the output signals.

3.2.1.1 Cone and Test Vector Generation

As evidenced by the examples, the Sogomonyan algorithm was devised essentially for the partitioning of combinational circuits into cones, which is well known in the literature and used to reduce labor tediousness during test vector generation [2–4].

Following the partitioning, a definite group of circuit primary inputs lies in the base of each cone. It is only the cone base that defines the performance result at the cone apex. In this way the generated cones can be made independent of one another.

The cone is the additional circuit logical characteristic that enables efficient troubleshooting inside the circuit and transmission of relevant messages to a circuit primary output that serves as the cone apex. The number of circuit primary outputs determines the number of its cones (Figure 3.4). Some of the primary inputs affect various primary outputs concurrently, which makes the intersection of cone bases possible (Figure 3.5).

Partitioning of a large circuit into cones permits the user to generate tests for each cone separately (Figure 3.6), considerably reducing the number of necessary test vectors. For instance, the circuit depicted in Figure 3.4 has 14 primary inputs (In1, In2, In3, In4, In5, In6, In7, In8, In9, In10, In11, In12, In13, In14) and 3 primary outputs that are the apices of three similarly named cones (Out1, Out2, Out3). The Out1 cone has eight primary In1–In8 inputs and contains ten gates (G1, G2, G3, G4, G9, G10, G11, G12, G14, G15, G16, G18, G19). The Out2 cone has six primary In5–In10 inputs and contains seven gates (G3, G4, G5, G11, G12, G16, G19). The Out3 cone has six primary In9–In14 inputs and also contains seven gates (G5, G6, G7, G8, G13, G17, G20).

For the sake of simplicity, we will consider the set of constant faults at the outputs of F1–F20 gates (specified faults of a fault universe). For the worst

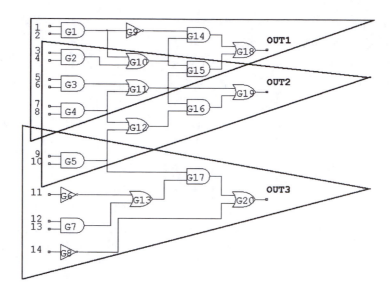

FIGURE 3.4
Cone definition.

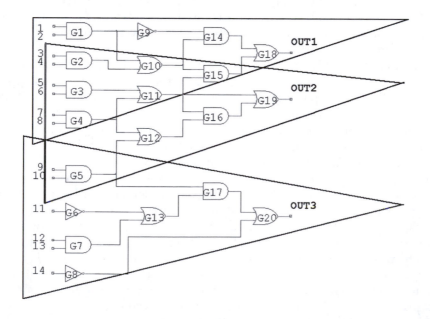

FIGURE 3.5
Cone intersection.

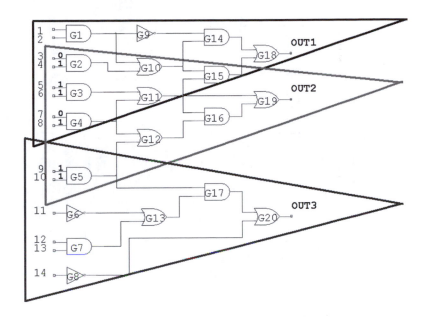

FIGURE 3.6
Test generation for cones.

case, we can calculate the number of test vectors required for the refinement of all faults being studied:

N1(In1–In14 | Out1+Out2+Out3 | F1–F20)
   = N(In1–In8 | Out1 | F1, F2, F3, F4, F9, F10, F11, F12, F14, F15, F16, F18, F19)
      + N(In5–In10 | Out2 | F3, F4, F5, F11, F12, F16, F19)
      + N(In9–In14 | Out3 | F5, F6, F7, F8, F13, F17, F20)
   = $2^8 \times 10 + 2^6 \times 7 + 2^6 \times 7 = 3456$.

When the circuit is not partitioned into cones, we need 327,680 test vectors for the worst vectors:

N2(In1–In14 | Out1+Out2+Out3 | F1–F20) = $2^{14} \times 20 = 327680$.

Thus, we have a nearly 99% reduction of test vectors for said circuit with preferentially horizontal links between the circuit elements, from inputs to outputs:

$$1 - 3456/327680 = 0.98945.$$

Certainly, by circuit partitioning into cones even a minor number of vertical links in the circuit results in an efficiency decrease. Thus, the circuit given in Figure 3.7 has:

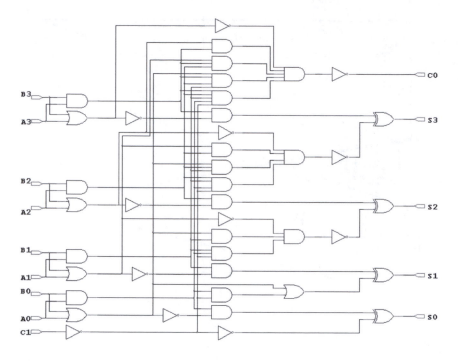

FIGURE 3.7
An example of a circuit with both horizontal and vertical links between the elements.

1. Nine primary inputs: CI, A0, B0, A1, B1, A2, B2, A3, and B3.
2. Five primary outputs that are the apices of five similarly named cones: S0, S1, S2, S3, and C0.
   a. Cone S0, whose base has 3 primary inputs, CI, A0, and B0, contains 7 gates: G1, G2, G3, G10, G14, G15, and G39.
   b. Cone S1, whose base has 5 primary inputs, CI, A0, B0 A1, and B1, contains 10 gates: G1, G2, G3, G4, G5, G11, G16, G17, G32, and G40.
   c. Cone S2, whose base has 7 primary inputs, CI, A0, B0, A1, B1, A2, and B2, contains 15 gates: G1, G2, G3, G4, G5, G6, G7, G12, G18, G19, G20, G21, G33, G36, and G41.
   d. Cone S3, whose base has 9 primary inputs, CI, A0, B0, A1, B1, A2, B2, A3, and B3, contains 18 gates: G1, G2, G3, G4, G5, G6, G7, G8, G9, G13, G22, G23, G24, G25, G26, G34, G37, and G42.
   e. Cone C0, whose base has the same 9 primary inputs as cone S3, CI, A0, B0, A1, B1, A2, B2, A3, and B3, contains 16 gates: G1, G2, G3, G4, G5, G6, G7, G8, G9, G27, G28, G29, G30, G31, G35, and G38.

Again, for the worst case, we can calculate the number of test vectors required for the refinement of all faults being studied, making allowances for the fact that some faults for the current cone were already present in the previous cones:

$N1(CI–C0 \mid S0+S1+S2+S3+CO \mid F1–F42)$
$= N(CI–B0 \mid S0 \mid F1, F2, F3, F10, F14, F15, F39)$
$+ N(CI-B2 \mid S1 \mid F4, F5, F11, F16, F17, F32, F40)$
$+ N(CI–B3 \mid S2 \mid F6, F7, F12, F18, F19, F20, F21, F33, F36, F41)$
$+ N (CI–B3 \mid S3 \mid F8, F9, F13, F22, F23, F24, F25, F26, F34, F37, F42)$
$+ N (CI–B3 \mid CO \mid F27, F28, F29, F30, F31, F35, F38)$
$= 2^3 \times 7 + 2^5 \times 7 + 2^7 \times 10 + 2^9 \times 11 + 2^9 \times 7 = 10776.$

If the circuit is not partitioned into cones, 21504 test vectors are necessary for the worst case:

$$N2(CI–CO \mid S0+S1+S2+S3+CO \mid F1–F42) = 2^9 \times 42 = 21504.$$

Thus, again we have a nearly 50% reduction of test vectors for said circuit with preferentially vertical links between its elements:

$$1 - 10776/21504 = 0.49889.$$

3.2.1.2 *Partitioning Circuits into Cones and Subcones Using Simulink®*

The techniques for partitioning circuits into cones in the Simulink environment have the advantage of high visibility, which is extremely helpful for the test designer. Figure 3.8 represents a model in the Simulink environment

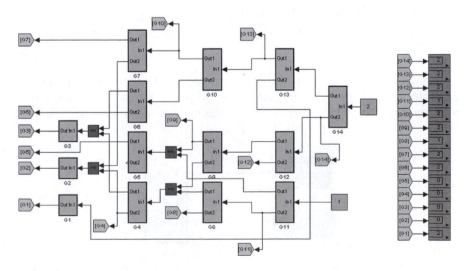

FIGURE 3.8
Cone definition model in Simulink.

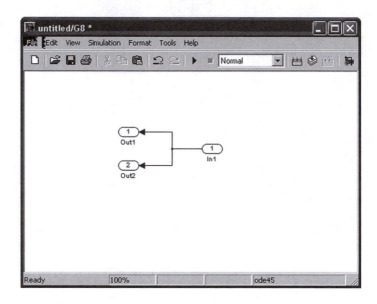

FIGURE 3.9
A gate model for a cone definition model in Simulink.

for the circuit to be partitioned into cones (Figure 3.4). The model has the following constituents:

1. A model of an n-input logical gate (Figure 3.9)
2. A primary input model of the circuit (Figure 3.10)
3. A primary output model (Figure 3.11)
4. A circuit gate indication unit found inside the constructed cone (Figure 3.12).

Cone construction in the model environment is initiated by setting a certain value (0, 1, 2,...) for the Constant Block, which is the model of the circuit primary output, that is, the corresponding cone apex, and consequent simulation

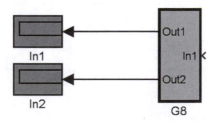

FIGURE 3.10
A primary input model for a cone definition model in Simulink.

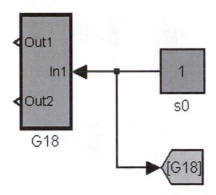

FIGURE 3.11
A primary output model for a cone definition model in Simulink.

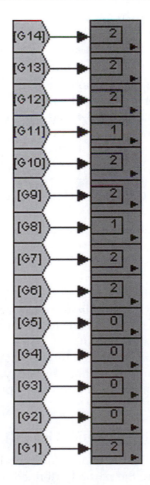

FIGURE 3.12
A cone gates indication for a cone definition model in Simulink.

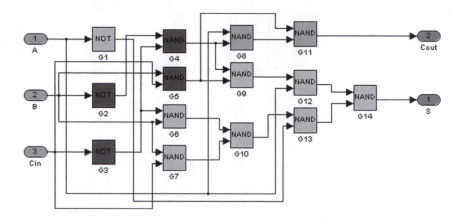

FIGURE 3.13
The result of partitioning a circuit into cones.

startup. The cone construction result is initiated by means of a Display Block, which is the model of the circuit primary inputs and the indication unit for the gates found inside the constructed cone.

Our (the adder circuit in Figure 3.4; see also Figure 3.13) example shows that the circuit is partitioned into two cones:

$$1 \to G8, G11,$$

$$2 \to G1, G6, G7, G9, G10, G12, G13, G14,$$

whereas the behavior of other four gates,

$$0 \to G2, G3, G4, G5,$$

influences both circuit outputs (Figure 3.13).

The gates that influence various circuit outputs are detected using a Relational Operator Block; when various cone labels (in our case, 1 and 2) enter the unit inputs, zero emerges at its output.

The subcone construction is initiated by a Constant Block connection (constant fault model) to the output point of a respective logical gate found inside a certain cone and consequent simulation startup (Figure 3.15).

3.2.2 Fault Simulation for Sequential Circuits

For any sequential circuit with a known fault model, a fault detection function is built by minimizing the partially defined machine [1]. The notion of a fault detection circuit for a given sequential circuit and known fault model is introduced as follows.

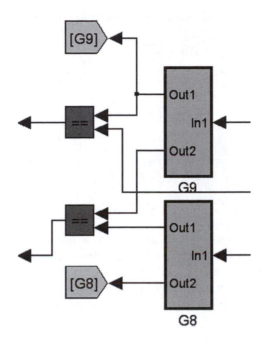

FIGURE 3.14
Detection of gates affecting various circuit outputs by means of a Relational Operator Block.

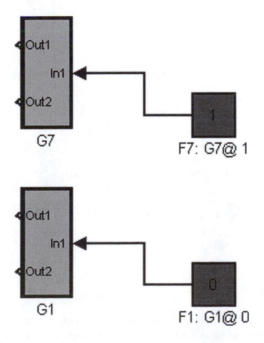

FIGURE 3.15
Initiation of subcone construction.

Let us name the machine-realizing sequential circuit A_s

$$A = (X, Y, Z, z_0, \delta, \lambda),$$

and let

$$F(A) = \{A_0, A_1, ..., A_n\},$$

where

$$A = A_0,$$

$$A_i = (X, Y_i, Z_i, z_{0i}, \delta_i, \lambda_i)$$

is a fault model for an A machine (and for an A_s sequential circuit, respectively). Then $A_1, ..., A_n$ are the machines that can be implemented via sequential circuits instead of A_0 machine, owing to the faults found in a given sequential circuit. In this process the A_0 machine is called a normal machine, and the sequential circuit that implements it is the fault-free circuit.

Example 3 [1]

Let the machine

$$A = (X = \{0,1\},$$

$$Y = Z = \{00, 01, 10, 11\}, \delta = \lambda),$$

$$\delta = (\delta_1, \delta_2),$$

$$\lambda = (\lambda_1, \lambda_2))$$

be implemented, and

$$y_1(t) = z_1(t + 1) = \delta_1(z_1(t), z_2(t), x(t)) = \lambda_1(z_1(t), z_2(t), x(t)) = \overline{x}(t) + z_1(t)^* \, z_2(t) \, ;$$

$$y_2(t) = z_2(t + 1) = \delta_2(z_1(t), z_2(t), x(t)) = \lambda_2(z_1(t), z_2(t), x(t)) = x(t)^* \, \overline{z}_1(t) + \overline{x}(t)^* \, z_2(t).$$

The machine's corresponding scheme is given in Figure 3.16.

In this process the DELTA combinational section is implemented via the functions described in Table 3.9.

The state diagram of the A machine is given in Figure 3.17, and the DELTA combinational section scheme, in Figure 3.18. Let the A machine fault model contain only one fault expressed as the absence of transition from state 01 to state 11 on the condition that input signal X possesses the value of 0 (see the state diagram in Figure 3.19).

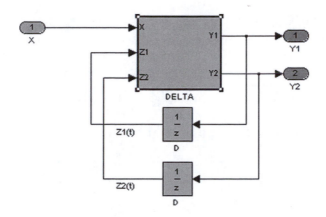

FIGURE 3.16
The scheme of the A machine to be implemented.

The simplest troubleshooting scheme is built on the duplication and comparison technique (Figure 3.20). The corresponding Simulink model is composed of:

1. The Signal Builder Block (X-TEST STIMULUS): The source of the troubleshooting test (the test-signal time diagram is shown in Figure 3.21).
2. The Chart Block of the Stateflow (CUT [circuit under test]: MEALY MACHINE): The tested object model of the A machine.
3. The Chart Block of the Stateflow (GOOD CIRCUIT: MEALY MACHINE): The reference object model — the reference fault-free A machine, with the state diagram shown in Figure 3.17.
4. The Subsystem Block (COMPARATOR): A comparator model that compares the tested object response with the reference response to the same input impact (the comparator inner structure is shown in Figure 3.22).
5. The Display Block (RESULT) for the simulation result mapping:
   0: The tested object does not have a fault (i.e., the state diagram of the CUT: Mealy Machine unit coincides with the reference state diagram of Figure 3.17.
   1: The tested object has the above-described fault (the state diagram of the CUT: Mealy Machine unit coincides with the Figure 3.19 diagram).
6. The Stop Simulation Block: Stops the simulation as soon as the fault is detected (discrepancies between the tested object responses and the reference response) by the comparator.

TABLE 3.9

Description of the z_1, z_2 Functions

X	z_1, z_2			
	00	01	10	11
0	00	11	00	11
1	01	01	00	10

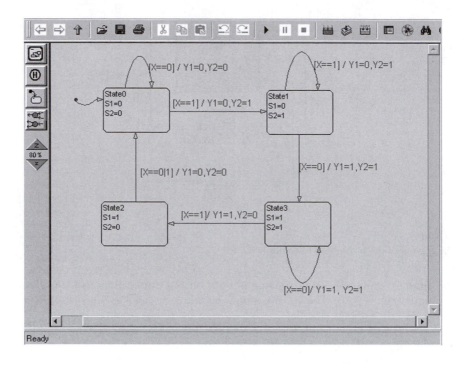

FIGURE 3.17
The state diagram of the A machine to be implemented.

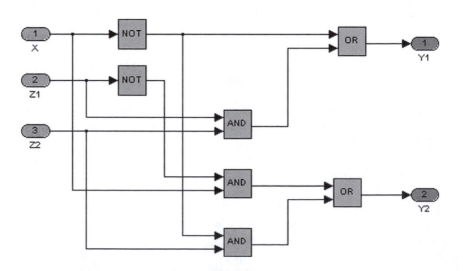

FIGURE 3.18
The DELTA combinational section scheme.

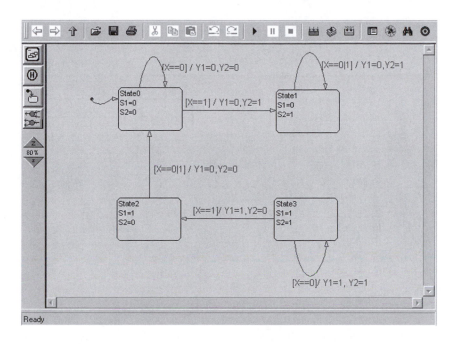

FIGURE 3.19
The state diagram of the fault-containing A_1 machine (absence of transition between states 01 and 11 states).

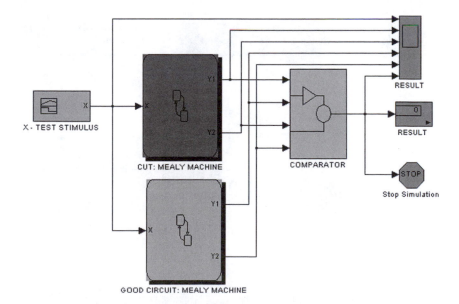

FIGURE 3.20
Simulink model of the troubleshooting scheme (the A machine model contains one fault) based on the duplication and comparison technique.

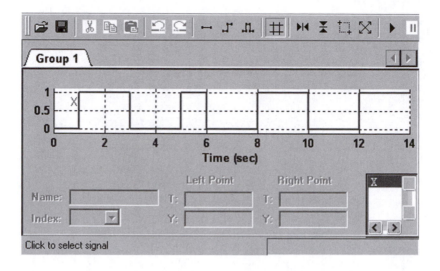

FIGURE 3.21
X test impact within the 0 to 14 simulation time units range.

7. The Scope Block (RESULT): Maps the testing as corresponding time diagrams. The Figure 3.23A time diagram reflects such testing that demonstrates that the tested object is in running order; the Figure 3.23B time
   diagram reflects such testing that demonstrates that the tested object contains the fault that can be detected at time t = 3.

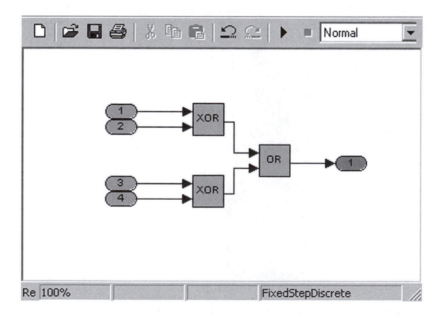

FIGURE 3.22
Internal structure of the COMPARATOR unit.

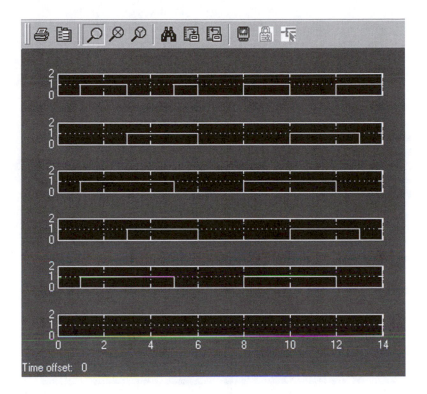

FIGURE 3.23A
Time diagram for the operable object testing.

Now we assume that instead of the A machine, one of the A_i $(i = 1,...,5)$ machines is implemented as a consequence of any possible fault. The Simulink model of the duplication and comparison scheme for this case is shown in Figure 3.24.

We proceed from the duplication and comparison schemes to a more optimal scenario — the troubleshooting machine. The $L = (X \times Y, \{0,1\}, Z_L, z_{0L}, \delta_L, \lambda_L)$ machine with input $X \times Y$ and output $\{0,1\}$ alphabets is called the fault troubleshooting machine for the A machine with the $F(A)$ fault model if the following condition is met for all $x_1, x_2,..., x_n \in X' \subseteq X$:

$$\lambda_L (z_{0L}, x_1 y_1 \cdots x_n y_n) = 0 - \text{ for } y_1 = \lambda(z_0, x_1), \quad y_2 = \lambda(z_0, x_1, x_2),$$

$$.....................$$ (3.1)

$$y_n = \lambda(z_0, x_1, x_2,..., x_n); \quad 1 - \text{ for } y_1 = \lambda(z_0, x_1), \quad y_2 = \lambda(z_0, x_1, x_2),$$

$$....................$$

$$y_{n-1} = \lambda(z_0, x_1, x_2,..., x_{n-1}) \quad \text{and} \quad \exists A_i \in F(A), \text{ where}$$

$$y_n = \lambda_i (z_0, x_1, x_2,..., x_n) \neq \lambda (z_0, x_1, x_2,..., x_n);$$

This macine is underdefined for all other cases.

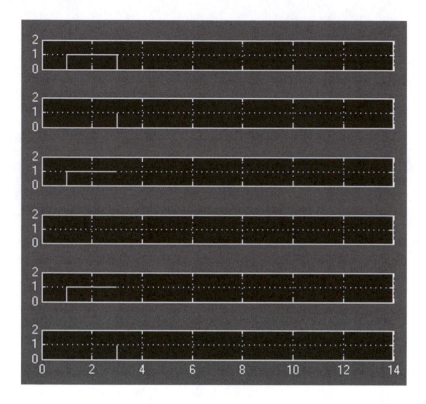

FIGURE 3.23B

Time diagram for testing the object that contains the fault indicated in Figure 3.19 and detected at time t = 3.

The L_s scheme that implements the L machine is called the troubleshooting scheme for the A machine. The presence of undefined values in Equation 3.1 results in the detection of faults only out of the fault model F(A) of sequential circuit L_s, whereas at the output of the troubleshooting scheme L_s, any value can be found at the moment of detection of the fault that is outside our fault model. The Equation 3.1 undefined values are used for scheme optimization.

If a certain fault model is located and the only known fact is that the A_s combinational circuit can contain only one of these faults, then for $x_1, x_2,...,$ $x_n \in X' \subseteq X$, instead of Equation 3.1, the following condition is met:

$$\lambda_L (z_{0L}, x_1y_{1...} x_ny_n) = 0 - \text{for } y_1 = \lambda(z_0, x_1), \quad y_2 = \lambda(z_0, x_1, x_2),$$

$$\cdots\cdots\cdots\cdots\cdots$$

$$y_n = \lambda(z_0, x_1, x_2,..., x_n); \quad 1 - \text{for } y_1 = \lambda(z_0, x_1), y_2 = \lambda(z_0, x_1, x_2), \qquad (3.2)$$

$$\cdots\cdots\cdots\cdots\cdots$$

$$y_{n-1} = \lambda(z_0, x_1, x_2,..., x_{n-1})$$

$$y_n \neq \lambda (z_0, x_1, x_2,..., x_n);$$

This machine is underdefined for all other cases.

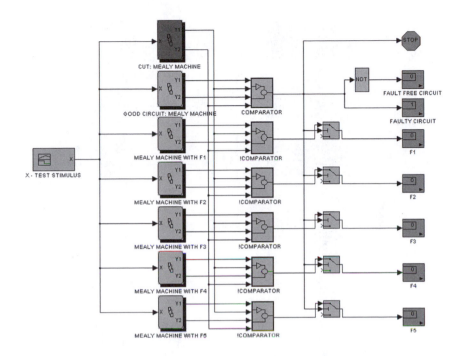

FIGURE 3.24
The duplication and comparison scheme for the machine whose fault model contains five possible faults.

Direct implementation of the troubleshooting machine L_s defined from Equation 3.1 for any finite A machine with a fault model of

$$F(A) = \{A_0, A_1,..., A_n\}, (n > 0)$$

is given in Figure 3.25.

The troubleshooting scheme is composed of the following components:

1. The $A = A_0, A_1,..., A_n$ machines for the fault model
2. $n + 1$ comparison schemes
3. A combinational circuit that implements the partially defined Boolean function f
4. A D flip-flop and OR gate

Input signals of the troubleshooting machine X are given to all fault model machines simultaneously. Output signals of $A_0, A_1,..., A_n$ machines are compared (in relevant comparison schemes) with the Y output signal of sequential circuit A_s to be controlled. If both input signals of the comparison scheme coincide, a null is generated at the output of each u_i $(1 \leq i \leq n)$ comparison scheme, and a unity is generated if they do not coincide.

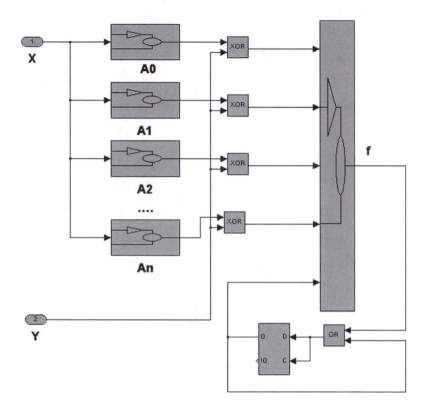

FIGURE 3.25
The troubleshooting scheme L_s.

The combinational circuit output f is connected to the combinational circuit $(n + 2)$th input f via a gate OR and D flip-flop. The D flip-flop initial state is zero. When the combinational circuit f generates its first unity, the D flip-flop passes to the unity state, which, owing to the available feedback, will be preserved at the combinational circuit $(n + 2)$th input as well.

The following relation is held for the f function:

$$0 \text{ for } u_0 = 0, u_{n+1} = 0;$$

$$1 \text{ for } u_0 = 1, u_1 = 0, u_{n+1} = 0;$$

$$1 \text{ for } u_0 = 1, u_2 = 0, u_{n+1} = 0; \tag{3.3}$$

$$f(u_0 \, u_1, \ldots, u_{n+1}) = \ldots\ldots\ldots\ldots\ldots$$

$$1 \text{ for } u_0 = 1, u_n = 0, u_{n+1} = 0;$$

Underdefined values result for all other cases.

The L machine output becomes undefined after the first unity, and these undefined values are used for machine optimization.

A partially defined L machine possesses $m_0 \times m_1 \times \dots \times m_n \times 2$ states, where m_i is the number of states of the A_i ($i = 0, 1, \dots, n$) machine. For instance, for $n = 10$ $m_0 = m_1 = \dots = m_{10}$, 2×10^{11} possible states can be obtained, and therefore, none of the known optimization techniques for finite automata can be applied for this combinational circuit. However, a reduced L_{red} machine with a maximally decreased number of states possesses only m_0 states.

The f undefined values in Equation 3.3 can be redefined as follows:

$$f(u_0\, u_1, \dots, u_{n+1} = \begin{cases} 0 \text{ for } u_0 = 0; \\ 1 \text{ for } u_0 = 1, \quad u_1 = u_2 = \dots = u_n = 1; \end{cases} \tag{3.4}$$

We arrive at the relation

$$f(u_0\, u_1, \dots, u_{n+1}) = \begin{cases} 0 \text{ for } u_0 = 0; \\ 1 \text{ for } u_0 = 1; \end{cases} \tag{3.5}$$

In these conditions, the troubleshooting scheme given in Figure 3.25 is similar to duplication and comparison and is constructed from the L_{red} reduced machine and the comparison scheme.

To specify the reduced machine, the starting task will be to transform the $A = A_0 = (X, Y, Z, z_0, \delta, \lambda)$ normal machine into the partially defined $L' = (X \times Y, \{0,1\}, Z_L, z_{0L}, \delta_L, \lambda_L)$ machine that is L equivalent and possesses as many states as A. The partially defined L' machine is minimized using a conventional number-of-states minimization technique for the partially defined machines.

The A machine transformation into the L' machine is made in four steps.

Step 1:
For $x \in X$; $y \in Y$; $z, z' \in Z$ is given
$\delta_L'(z, (x, y)) = z'$ for $\delta(z, x) = z'$ and $\lambda(z, x) = y$;
$\lambda_L'(z, (x, y)) = 0$ for $\lambda(z, x) = y$.

In this process, the arc in the A machine state diagram, leading from z to z' and specified by the x, y parameters, is replaced by the arc specified as (x, y),0. Such an L' machine does not detect any fault until the tested machine behaves as an operable $A_0 = A$ machine.

Step 2: For each A machine ∈ F(A), a G_i fault-detecting tree is given that describes the difference between the A and A_i states up to the first output-detected fault.

Step 3: For all z ∈ Z, x ∈ X, y ∈ Y whose λ'_L and δ'_L functions are still undefined after the first two steps, the undefined states are set, and a labeled (x, y) arc, having a free final point, is introduced into the state diagram.

Step 4: A partially defined L' machine with m = m_0 states is minimized by any accepted technique.

The approach described is exemplified in Figure 3.17 by the A machine state diagram. Let us assume that owing to potential faults, one of the A_i machines, where i = 1,..., 16, will be implemented. In this case the fault model looks as follows:

$$F(A) = \{A_0 = A, A_1, A_2, ..., A_{16}\}.$$

Each of $A_1, A_2, ..., A_{16}$ machines has input and output sets, sets of states, and initial state 00 identical to those of the A machine. The $A_i = (X, Y \in Z, z_0, \delta_i, \lambda_i)$ machines differ from the A machine in their δ_i, λ_1 functions with regard to the fact that the differences in the changes of state are displayed in one order only. Thus, for instance, the A_1 machine differs from the A machine in that the A_1, which is in state 00, passes to state 01 by arriving at input 0, while the A machine remains in state 00 (states 00 and 01 differ by one order) (Figure 3.26).

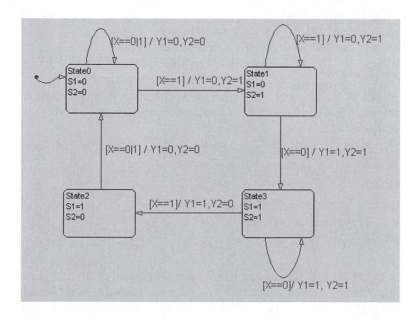

FIGURE 3.26
The state diagram of the A_1 machine.

For all pairs of (A, A_i), where i = 1, 2,..., 16, the G_i fault diagnostic trees are built. For example, let us examine the construction of the G_1 fault diagnostic tree related to the (A, A_1) pair. In this process we will encode states 00, 01, 10, and 11 and output signals as 0, 1, 2, 3 and use the following definitions: The tree root is depicted as an oval, the first-order finite node as a circumference, the second-order finite node as a rectangle, and a common node as a square.

The G_1 fault diagnostic tree is shown in Figure 3.27. The figure shows that the G_1 tree root equals (0,0) because state 0 is the initial state both for A and for A_1 machines (the encoded form of 00). This root is oval-enclosed.

With zero input, the A machine generates the $\lambda(0,0) = 0$ output signal and passes to state 0, whereas the A_1 machine, with 0 input, generates the $\lambda(0,1) = 1$ output signal and passes to state 1. In line with this, the 0(01) arc leads from the (0,0) node to the first-order finite node (0,1) denoted by the circumference. No arc leaves the node.

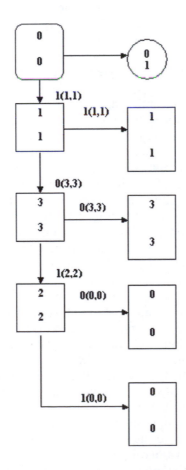

FIGURE 3.27
The G_1 fault diagnostic tree.

With one input in state 1, the A machine, similarly to the A_1 machine, passes to state 1 and generates the output signal equal to 1. The corresponding arc in the the G_1 tree is indicated by the (11) unit and runs from the (0,0) node to the (1,1) node, indicated by the square, like all usual nodes. When 1 enters state 1, the A machine, similarly to the A_1 machine, passes to state 1 and generates 1. Therefore, the arc labeled 1(11) in the G_1 tree leads from the (1,1) node to another (1,1) node, the finite second-order node, because the (1,1) node has already been presented (as the preceding node); it is denoted with a rectangle, and no outgoing arcs are found.

The 0(3,3) arc runs from the common node (1,1) to the common node (3,3) because the $\delta(1,0) = \delta_1(1,0) = 3$, $\lambda(1,0) = \lambda_1(1,0) = 3$ conditions are met. The 0(3,3) arc runs from the common node (3,3) to another node (3,3), the finite second-order node, because the $\delta(3,0) = \delta_1(3,0) = 3$, $\lambda(3,0) = \lambda_1(3,0) = 3$ conditions are met. Moreover, the 1(2,2) arc runs from the common node (3,3) to the common node (2,2), because $\delta(3,1) = \delta_1(3,1) = 2$, $\lambda(3,1) = \lambda_1(3,1) = 2$.

The 0(0,0) arc in the G_1 tree runs from the common node (2,2) to the (0,0) node, the finite second-order node, because the $\delta(2,0) = \delta_1(2,0) = 0$, $\lambda(2,0) = \lambda_1(2,0) = 0$ conditions are met and because the (0,0) node has already occurred (as the G_1 root). Finally, the 1(0,0) arc runs from the common node (2,2) to yet another finite second-order node (0,0), because the $\delta(2,1) = \delta_1(2,1) = 0$, $\lambda(2,1) = \lambda_1(2,1) = 0$ conditions are met.

It can be seen that the G_1 fault diagnostic tree describes the first emergence of an operational difference between the A and A_1 machines using the finite first-order node (in this case, the [0,1] node). If an arbitrary input sequence is applied, the G_1 tree assists in detecting the moment when the error that converts the A machine into the A_1 machine first appears at the scheme output. The G_2, G_3,..., G_{16} fault diagnostic trees are defined similarly.

Later on, a partially defined L′ machine that has as many states as the A machine and is equivalent to the L troubleshooting machine is defined by the modification of an operable $A = A_0$ machine. According to the first step of the algorithm, the arc in the A machine state diagram that leads from z to z′ and is labeled (x,y) is replaced by the (x,y),0 labeled arc (Figure 3.28). Thus, we ensure that no fault signal will be generated when the controlled scheme operates regularly. To further define the L′ machine, all nodes preceeding the finite first-order nodes should be found in the G_1, G_2,..., G_{16} fault diagnostic trees. If (z_1, z_2) is a node proceeding the finite first-order nodes $(\delta[z_1, x], \delta_i[z_2, x])$, then in the L′ graph an arc labeled $(x, \lambda_i[z_2, x])$ should be drawn so that its second end remains free.

The G_1 fault diagnostic tree contains the finite first-order node $(0,1) = (\delta[0, 0], \delta[0, 0])$ with the preceeding (0,0) node. Therefore, when graph L′ is in state 0, the arc labeled $(0, \lambda[0, 0])$, $1 = (0,1)1$ should be drawn so that its second end remains free (an indefinite value). The L′ = $(X*Y, \{0,1\}, Z, z_0, \delta_{L'}, \lambda_{L'})$ machine possesses four states, similar to the operable A machine. The L′ machine graph is described by $\delta_{L'}$, $\lambda_{L'}$ functions in Table 3.10.

Finally, by reducing the number of states in the L′ machine, we obtain the L troubleshooting machine, and having implemented it, we see the desired

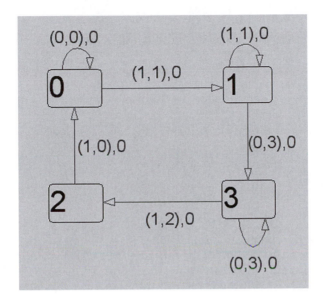

FIGURE 3.28
Modified state diagram of the A machine.

TABLE 3.10

The L′ Machine Graph Description

z	xy	$\chi_{L'}(z, xy)$	$\kappa_{L'}(z, xy)$
0	00	0	0
0	01	—	1
0	02	—	1
0	03	—	—
0	10	—	1
0	11	1	0
0	12	—	—
0	13	—	1
1	00	—	—
1	01	—	1
1	02	—	1
1	03	3	0
1	10	—	1
1	11	1	0
1	12	—	—
1	13	—	1
2	00	0	0
2	01	—	1
2	02	—	1
2	03	—	—
2	10	0	0
2	11	—	1
2	12	—	1

(continued)

TABLE 3.10

The L′ Machine Graph Description (Continued)

z	xy	$\chi_{L'}(z, xy)$	$\kappa_{L'}(z, xy)$
2	13	—	—
3	00	—	—
3	01	—	1
3	02	—	1
3	03	3	0
3	10	—	1
3	11	—	—
3	12	2	0
3	13	—	1

troubleshooting scheme. As long as states 0, 1, and 3 of the L′ machine are compatible, they can be pooled to form the A = (0,1,3) compatibility class. State 2 is incompatible with any other state and forms a compatibility class of its own, B = (2). With this in mind, the following representation can be obtained from the previous table: L = (X*Y, {0,1}, {A,B}, A, δ_L, λ_L machine for δ_L, λ_L (Table 3.11).

It is evident that the L troubleshooting scheme has only two states, whereas the controlled A machine has four states. If we encode the A = 0, B = 1 and y = 0,1,2,3 values as 00, 01, 10, 11, then for the L_S troubleshooting scheme we have the implementation with tabulated δ_L, λ_L. This implementation is shown in Figure 3.29. The inner structures of the DELTA and LAMBDA units are shown in Figure 3.30 and Figure 3.31, respectively. The Chart Block of the Stateflow (CUT: Mealy Machine) is the tested object — the A machine — model.

TABLE 3.11

Optimized L′ Machine Graph Description

z	xy	$\delta_L(z, xy)$	$\lambda_L(z, xy)$
A	00	A	0
A	01	—	1
A	02	—	1
A	03	A	0
A	10	—	1
A	11	A	0
A	12	B	0
A	13	—	1
B	00	A	0
B	01	—	1
B	02	—	1
B	03	—	—
B	10	A	0
B	11	—	1
B	12	—	1
B	13	—	—

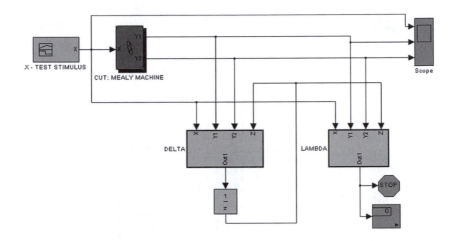

FIGURE 3.29
Implementation of the troubleshooting scheme L_s.

We will present an alternative approach to fault simulation for sequential circuits. Figure 3.32 shows the Simulink model for the determination of the potential number of paths (state change sequences) from Figure 2.31 algorithmic state machine (ASM) (Chapter 2). Each ASM state is represented by a Subsystem Block (A SOURCE and B, C, D, E, F, and A SINK units; their internal structures are shown in Figure 3.32A–G respectively). Only the starting state A is represented by two such units (A SOURCE and A SINK). The Constant Block, incorporated into the A SOURCE unit, is the signal source

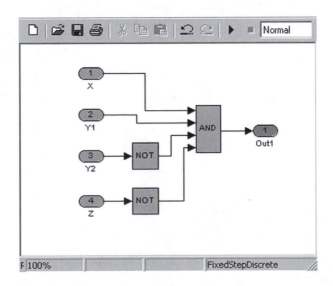

FIGURE 3.30
Internal structure of the DELTA unit.

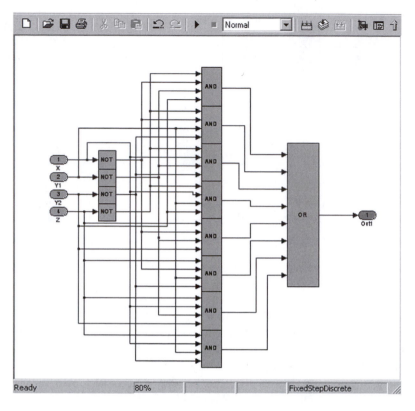

FIGURE 3.31
Internal structure of the LAMBDA unit.

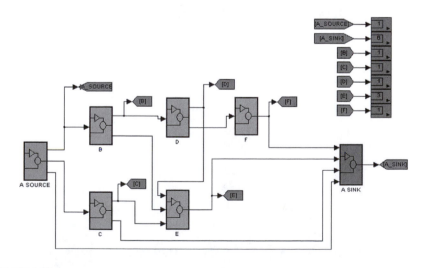

FIGURE 3.32
Simulink model for the determination of potential number of paths (state change sequences) from Figure 2.31 ASM (Chapter 2).

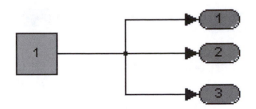

FIGURE 3.32A
Internal structure of the A SOURCE unit of the Simulink model.

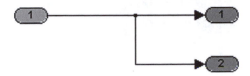

FIGURE 3.32B
Internal structure of the B unit of the Simulink model.

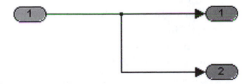

FIGURE 3.32C
Internal structure of the C unit of the Simulink model.

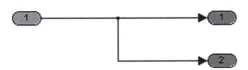

FIGURE 3.32D
Internal structure of the D unit of the Simulink model.

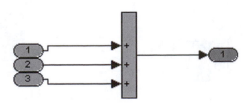

FIGURE 3.32E
Internal structure of the E unit of the Simulink model.

FIGURE 3.32F
Internal structure of the F unit of the Simulink model.

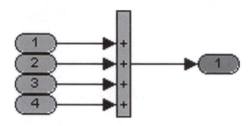

FIGURE 3.32G
Internal structure of the A SINK unit of the Simulink model.

in the model. It can be seen that if several elementary linkpaths lead to the same state, then the Sum Block for the summation of the number of potential linkpaths can be found inside the respective unit (Figure 3.32E, G).

In case only one linkpath leads to a certain state, and several linkpaths leave it, the structures of the corresponding units are extremely simple (Figure 3.32A, B, C, D, F). The model indication block is composed of a number of Goto Blocks and From Blocks, as well as Display Blocks. It can be observed that the ASM has six potential paths (the Display Block that corresponds to the A SINK unit shows six).

After the number of potential paths has been determined, they are numbered by the Simulink model shown in Figure 3.33. The information propagates in the opposite direction, from the A SINK unit to the A SOURCE unit. The Constant Block, incorporated into the A SINK unit, is the signal source in this case (Figure 3.33A). The A SINK unit contains a list of potential

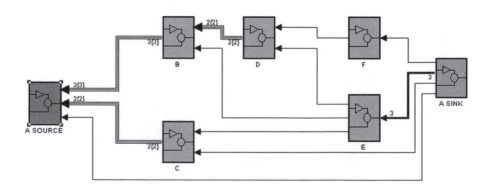

FIGURE 3.33
Simulink model for the numeration of paths from Figure 2.31 ASM (Chapter 2).

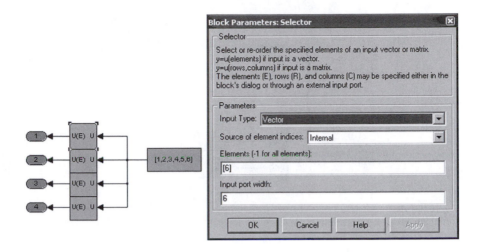

FIGURE 3.33A
Internal structure of the A SINK unit of the Simulink model.

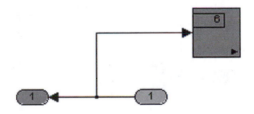

FIGURE 3.33B
Internal structure of the F unit of the Simulink model.

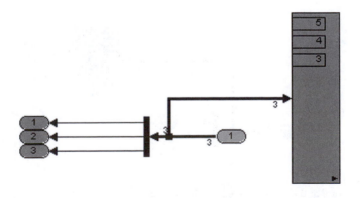

FIGURE 3.33C
Internal structure of the E unit of the Simulink model.

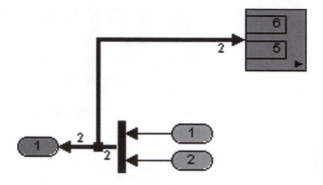

FIGURE 3.33D
Internal structure of the D unit of the Simulink model.

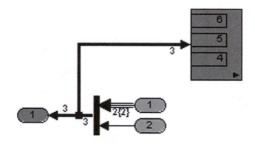

FIGURE 3.33E
Internal structure of the B unit of the Simulink model.

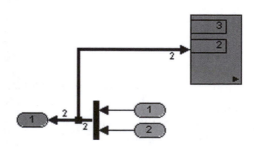

FIGURE 3.33F
Internal structure of the C unit of the Simulink model.

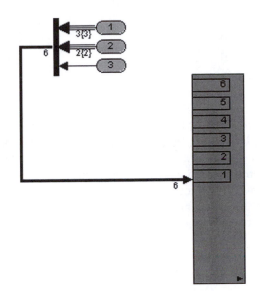

FIGURE 3.33G
Internal structure of the A SOURCE unit of the Simulink model.

numbered paths (1, 2, 3, 4, 5, 6; all six potential paths pass through the A state). The output of this Constant Block is connected to the inputs of four Selector Blocks (according to the number of paths leaving the A SINK unit). Figure 3.33A shows the Block Parameters dialog box for the upper Selector Block. Hence, it can be seen (the Elements box contains 6) that the path 6 number will arrive at the output of this Selector Block (it follows from Figure 3.32 that only one path leads to state A via state F). The content of the Elements box for the remaining three Selector Blocks is as follows:

5,4,3: If the numbers of 5,4,3 paths arrive at the output, then three paths pass through state E to state A.
2: If the number of path 2 arrives at the output, then only one path passes through state C to state A.
1: If the number of path 1 arrives at the output, then state A has one loop.

The F, E, D, B, C, and A SOURCE units assist in determining the path numbers that pass through each ASM state. Their internal structures are extremely simple (Figure 3.33B–G).

The data obtained, together with the data on the analyzed ASM (Chapter 2, Figure 2.27 and Figure 2.31), permit the user to infer what linkpaths are contained in each of the six potential paths:

1: L1
2: L3, L6

3: L3, L7, L10
4: L2, L4, L5, L10
5: L2, L4, L8, L10
6: L2, L4, L9, L11

It follows that we have six tests executed by the Figure 3.34 Simulink model to verify the Figure 2.31 ASM. The model is composed of six Signal Builder Blocks (L1, L3L6, L3L7L10, L2L4L5L10, L2L4L8L10, L2L4L9L11) that implement the six tests.

The first test validates the execution correctness for the L1 linkpath. The test is executed over one step so that the the X1, X2 input signals should have zero values, whereas the X3 value is of no importance (Figure 3.34A). The second test validates the correctness of successive execution of the L3 and L6 linkpaths. The test is executed over two steps. The first step's imperative condition is the X1 setting to the zero state and the X2 setting to unity state. During the second step X1 should remain in the zero state (Figure 3.34B). The third test validates the correctness of the successive execution of the L3, L7, and L10 linkpaths. The test is executed over three steps (Figure 3.34C).

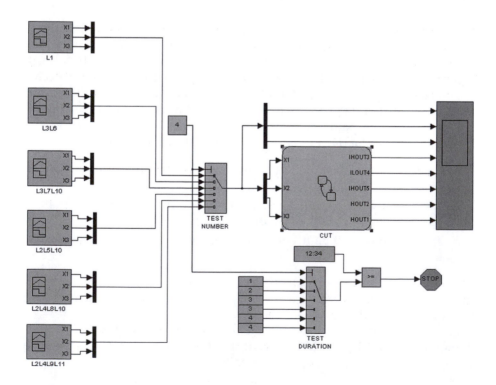

FIGURE 3.34
Test bench Simulink model for the Figure 2.31 ASM (version 1).

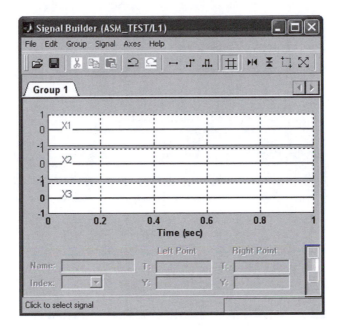

FIGURE 3.34A
Internal structure of the L1 unit of the test bench Simulink model.

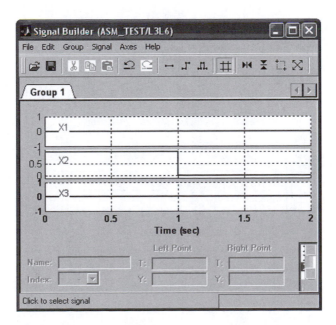

FIGURE 3.34B
Internal structure of the L3L6 unit of the test bench Simulink model.

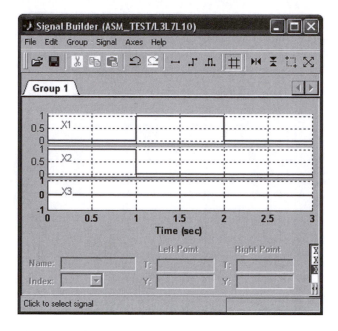

FIGURE 3.34C
Internal structure of the L3L7L10 unit of the test bench Simulink model.

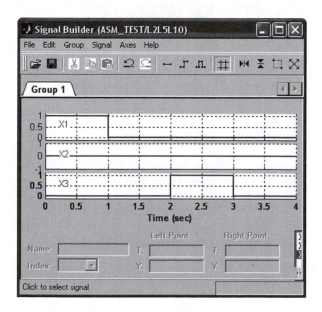

FIGURE 3.34D
Internal structure of the L2L5L10 unit of the test bench Simulink model.

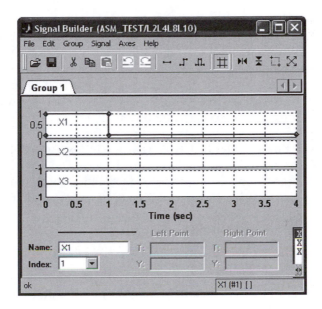

FIGURE 3.34E
Internal structure of the L2L4L8L10 unit of the test bench Simulink model.

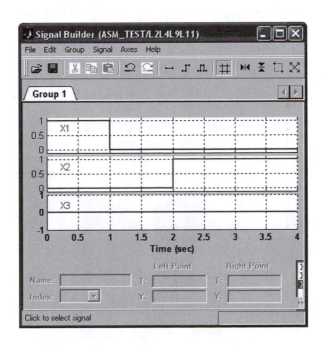

FIGURE 3.34F
Internal structure of the L2L4L9L11 unit of the test bench Simulink model.

Note that L10 is the unconditional transition, so the input signal states during the third step are of no importance. The fourth test validates the correctness of the successive execution of the L2, L4, L5, and L10 linkpaths. The test is executed over four steps (Figure 3.34D). The fifth test validates the correctness of the successive execution of the L2, L4, L8, and L10 linkpaths. The test is executed over four steps (Figure 3.34E). The sixth test validates the correctness of the successive execution of the L2, L4, L9, and L11 linkpaths. The test is executed over four steps (Figure 3.34F).

The Constant Block of Figure 3.34 assigns the index number of a requisite test (in this case test the number 4 has been chosen). The test number serves as the control signal for the Multiport Switch Block (with the name TEST NUMBER), which conducts the corresponding test stimuli to the circuit under test (CUT) inputs (Figure 2.31 ASM). The Display Block determines the execution correctness for the tested path.

The test execution time (the number of necessary steps) is specified by the following blocks:

1. Six Constant Blocks that contain the step number data for each test
2. The Multiport Switch Block TEST DURATION (its control signal is the same test number)
3. The Digital Clock Block, which is the simulation time source
4. The Relational Operator Block (> =) that yields a unity signal at the moment when all steps of a chosen test have been executed
5. The Stop Simulation Block that stops simulation at the point when a unity signal appears at the Relational Operator Block output

One more modification of the test bench Simulink model for the Figure 2.31 ASM is shown in Figure 3.35. The model includes the CUT and the fault-free circuit (FFC) for the Figure 2.31 ASM. Internal structures of TEST DURATION, TEST STIMULUS, and COMPARATOR units (Figure 3.35A, B, C, respectively) do not need any special comments in view of above-mentioned remarks concerning the Figure 3.34 test bench Simulink model. The TEST DURATION unit is intended for assigning the needed number of test steps.

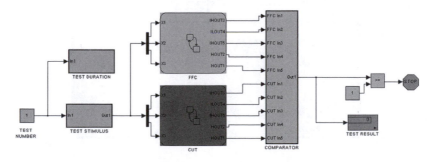

FIGURE 3.35
Test bench Simulink model for the Figure 2.31 ASM (version 2).

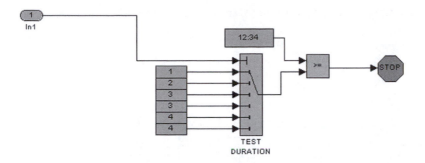

FIGURE 3.35A
Internal structure of the TEST DURATION unit of the test bench Simulink model.

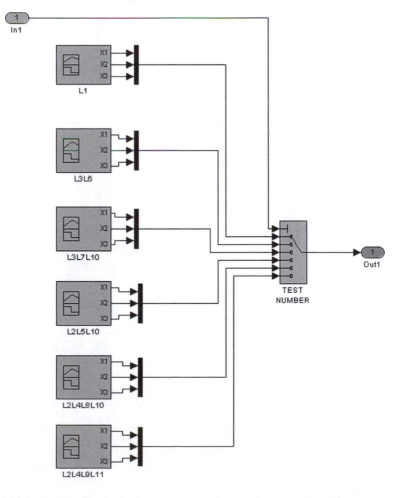

FIGURE 3.35B
Internal structure of the TEST STIMULUS unit of the test bench Simulink model.

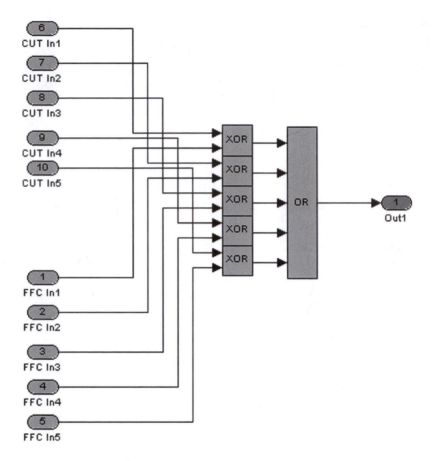

FIGURE 3.35C
Internal structure of the COMPARATOR unit of the test bench Simulink model.

The TEST STIMULUS unit sets corresponding test stimuli, whereas Constant Block called TEST NUMBER is intended for the required test selection.

References

1. Graf S., Gossel M. *Fehlererkennungsschaltungen* (in German). Akademie-Verlag, Berlin, 1987.
2. Abramovici M., Breuer M.A., Friedman A.D. *Digital Systems Testing and Testable Design.* IEEE Press, New York, 1995.
3. Bushnell M.L., Agrawal V.D. *Essentials of Electronic Testing for Digital, Memory & Mixed-Signal VLSI Circuits.* Kluwer Academic Publishers, Dordrecht, The Netherlands, 2004.
4. Navabi Z. *VHDL: Analysis and Modeling of Digital Systems.* McGraw-Hill, Singapore 1998.

4

Testability Analysis Methods

4.1 Combinational Controllability and Observability Analysis Models

At least two calculation options exist for the combinational controllability and combinational observability parameters in the combinational circuit nodes: (1) calculation of combinational controllability as the number of elementary operations necessary for setting the given circuit unit to logical zero or to logical unity, and calculation of combinational observability as the number of elementary operations necessary for the logical state observation in a given circuit unit representing the output of a certain gate [1–3]; and (2) calculation of CO as potential changes in the logical state of a unit following the changes in logical states at the circuit primary inputs and calculation of OB as potential changes in logical states of the circuit primary outputs, following the changes in logical states of a circuit node.

Making these calculations, we use the formulas that were first proposed in the SCOAP (Sandia Controllability/Observability Analysis Program). We cite these formulas for a number of gates and construct the corresponding Simulink® models [1–3]. First, we introduce the following definitions:

CC: Combinational controllability calculated as the number of elementary operations necessary for setting the circuit-specified unit to a certain logical state.

CC0: Combinational controllability calculated as the number of elementary operations necessary for setting the circuit-specified unit to logical zero (0-Combinational controllability).

CC1: Combinational controllability calculated as the number of elementary operations necessary for setting the circuit-specified unit to the logical unity (1-Combinational controllability).

OC: Combinational observability calculated as the number of elementary operations necessary for the logical state observation in a given circuit unit representing the output of a certain gate. The name of the circuit unit for which the testability parameter is calculated will be given in parenthesis.

4.1.1 The AND Gate

Out = AND(In1, In2).

$$CC0(Out) = \min\{CC0(In1), CC0(In2)\} + 1; \qquad (4.1)$$
$$CC1(Out) = CC1(In1) + CC1(In2) + 1; \qquad (4.2)$$
$$OC(In1) = CC1(In2) + OC(Out) + 1. \qquad (4.3)$$

The corresponding Simulink model (CC-AND) for the CC0, CC1 calculation is given in Figure 4.1. It has two inputs: CC(In1) and CC(In2), where the {CC0(In1), CC1(In1)} and {CC0(In2), CC1(In2)} vectors arrive, respectively, whereas the vectors of these parameters are recorded at the CC(Out) output for those gate inputs that are connected to the output, that is, {CC0(Out), CC1(Out)}. The model's internal structure is composed of two units, CC0-AND and CC1-AND, for the separate calculation of the CC0 and CC1 parameters. The structures of these units are very simple and are constructed from Equations 4.1 and 4.2. Two Demux Blocks "split" input vectors into scalars, and the Mux Block at the model output reduces the calculation results to the output vector shape.

Figure 4.2 gives a numerical example, which shows that if the logical zero is set at the In1 input following four elementary operations, {CC0(In1) = 4},

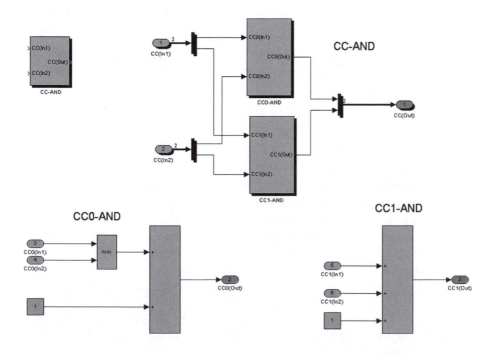

FIGURE 4.1
Simulink model of the AND gate for the CC0, CC1 calculation at its output from the known CC0, CC1 values at the gate inputs.

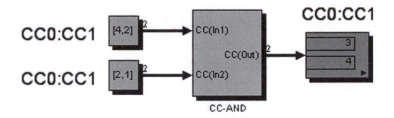

FIGURE 4.2
An example of CC0, CC1 calculations using the Simulink model of the AND gate.

and the logical zero is set at the In2 input following two elementary operations, {CC0(In2) = 2}, then at the gate output we have the controllability parameter value of CC0(Out) = 3, according to Equation 4.1. If the logical unity is set at the In1 input following two elementary operations, {CC1(In1) = 2}, and at the In2 input, following one elementary operation, {CC1(In2) = 1}, then the same state can be set at the output following four elementary operations, CC1(Out) = 4, according to Equation 4.2.

The Simulink model for the AND gate applied in the OC calculations is shown in Figure 4.3. The OC-AND model has three inputs, {OC(Out), CC1(In1), CC1(In2)}, and two outputs {OC(In1), OC(In2)}. Hence, it is evident that the OC parameter calculation flow for the circuit nodes streams out from right to left, from the primary outputs to the circuit primary inputs. It can clearly be seen that the model's inner structure is constructed in conformity with Equation 4.3. The corresponding numerical example is shown in Figure 4.4, and we see that the gate output's logical state is transferred to one of its inputs by performing a certain number of elementary operations sufficient for setting its alternative input to the logical unity (the Equation 4.3 formula). To calculate the OC parameter values for a gate input, we can use the CC1 parameter values for the rest of its inputs as well as the OC parameter value for the gate output.

4.1.2 The NAND Gate

Out = NAND(In1, In2).

$$CC0(Out) = CC1(In1) + CC1(In2) + 1; \tag{4.4}$$

$$CC1(Out) = \min\{CC0(In1), CC0(In2)\} + 1; \tag{4.5}$$

$$OC(In1) = CC1(In2) + OC(Out) + 1. \tag{4.6}$$

The corresponding Simulink model for the CC0, CC1 calculations is shown in Figure 4.5, and it is evident that the CC-NAND model structure is similar to the CC-AND model discussed above (Figure 4.1). The only difference is the inner structure of the CC0-NAND and CC1-NAND units constructed from Equations 4.4 and 4.5.

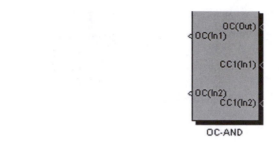

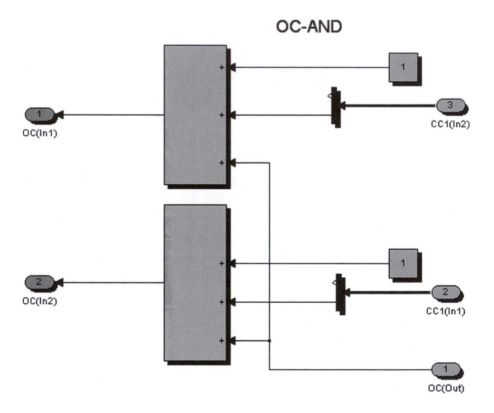

FIGURE 4.3
Simulink model of the AND gate for OC calculations.

The Figure 4.6 numerical example shows that if the logical unity is set at the In1 input following two elementary operations, {CC1(In1) = 2}, and at the In2 input following one elementary operation, {CC1(In2) = 1}, then the controllability parameter of CC1(Out) = 4 can be obtained at the gate output, according to Equation 4.4. If the logical zero is set at the In1 input following four elementary operations, {CC0(In1) = 4}, and at the In2 input following two elementary operations, {CC0(In2) = 2}, then at the gate output we have the logical unity following three elementary operations, CC0(Out) = 3, according to Equation 4.5.

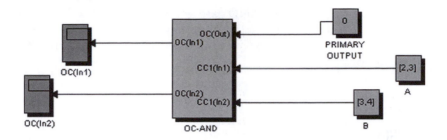

FIGURE 4.4
An example of OC calculations using the Simulink model of the AND gate.

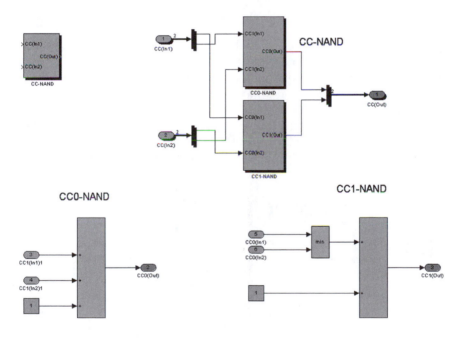

FIGURE 4.5
Simulink model of the NAND gate for the CC0, CC1 calculation at its output from the known CC0, CC1 values at the gate inputs.

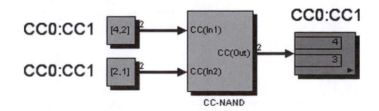

FIGURE 4.6
An example of CC0, CC1 calculations using the Simulink model of the NAND gate.

The Simulink model of the NAND gate applied in the OC calculations is similar to the Simulink model of the AND gate, shown in Figure 4.3, that is evidenced by the comparison of the corresponding Equation 4.3 and Equation 4.6 formulas.

4.1.3 The OR Gate

Out = OR(In1, In2).

$$CC0(Out) = CC0(In1) + CC0(In2) + 1; \tag{4.7}$$

$$CC1(Out) = \min\{CC1(In1), CC1(In2)\} + 1; \tag{4.8}$$

$$OC(In1) = CC0(In2) + OC(Out) + 1. \tag{4.9}$$

The corresponding Simulink model for the CC0, CC1 calculations is shown in Figure 4.7, and it is evident that the inner structures of the CC0 units is constructed according to Equations 4.7 and 4.8.

Figure 4.8 gives a numerical example, which shows that if the logical zero is set at the In1 input following four elementary operations, {CC0(In1) = 4}, and the logical zero is set at the In2 input following two elementary operations, {CC0(In2) = 2}, then at the gate output we have the controllability parameter value of CC0(Out) = 7, according to Equation 4.7. If the logical unity is set at

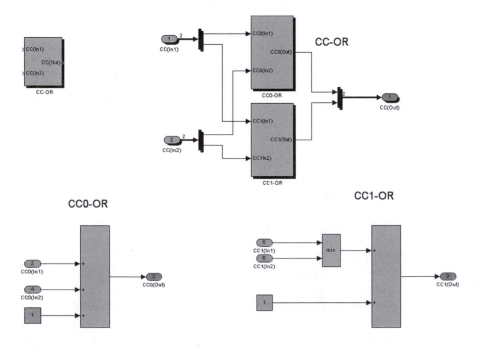

FIGURE 4.7
Simulink model of the OR gate for the CC0, CC1 calculation at its output from the known CC0, CC1 values at the gate inputs.

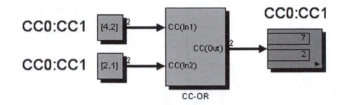

FIGURE 4.8
An example of CC0, CC1 calculations using the Simulink model of the OR gate.

the In1 input following two elementary operations, {CC1(In1) = 2}, and at the In2 input following one elementary operation, {CC1(In2) = 1}, then the logical unity can be set at the output following two elementary operations, CC1(Out) = 2, according to Equation 4.8.

The Simulink model of the OR gate applied in OC calculations is shown in Figure 4.9. The OC-OR model is constructed in conformity

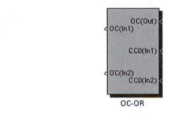

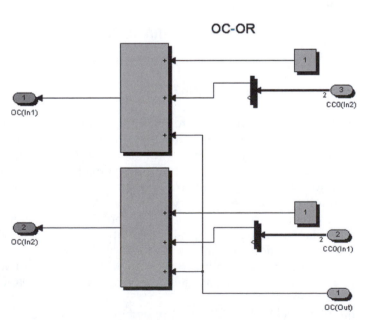

FIGURE 4.9
Simulink model of the OR gate for OC calculations.

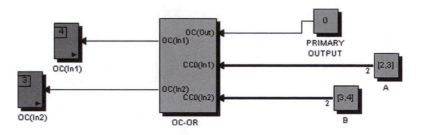

FIGURE 4.10
An example of OC calculations using the Simulink model of the OR gate.

with Equation 4.9. The corresponding numerical example is shown in Figure 4.10, and it is evident that the gate output logical state is transferred to one of its inputs by performing a certain number of elementary operations sufficient for setting its alternative input to the logical zero (the Equation 4.9 formula). To calculate the OC parameter values for a gate input, we can use the CC0 parameter values for the rest of its inputs as well as the OC parameter value for the gate output.

4.1.4 The NOR Gate

Out = NOR(In1, In2).

$$CC0(Out) = min\{CC1(In1), CC1(In2)\} + 1; \qquad (4.10)$$

$$CC1(Out) = CC0(In1) + CC0(In2) + 1; \qquad (4.11)$$

$$OC(In1) = CC0(In2) + OC(Out) + 1. \qquad (4.12)$$

The corresponding Simulink model for the CC0, CC1 calculations is shown in Figure 4.11, and it is evident that the inner node structures of the CC-NOR model are constructed according to Equations 4.10 and 4.11.

The Figure 4.12 numerical example shows that if the logical unity is set at the In1 input following two elementary operations, {CC1(In1) = 2}, and at the In2 input following one elementary operation, {CC1(In2) = 1}, then the controllability parameter of CC1(Out) = 2 can be obtained at the gate output, according to Equation 4.10. If the logical zero is set at the In1 input following four elementary operations, {CC0(In1) = 4}, and at the In2 input following one elementary operation, {CC0(In2) = 2}, then at the gate output we have the logical unity following seven elementary operations, CC0(Out) = 7, according to Equation 4.11.

The Simulink model of the NOR gate applied in OC calculations is similar to the Simulink model of the OR gate shown in Figure 4.9, as evidenced by the comparison of the corresponding Equations 4.9 and 4.12.

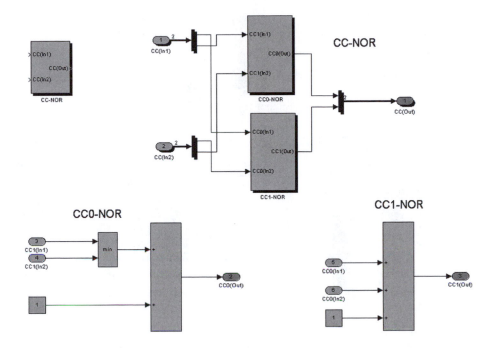

FIGURE 4.11
Simulink model of the NOR gate for the CC0, CC1 calculation at its output from the known CC0, CC1 values at the gate inputs.

4.1.5 The XOR Gate

Out = XOR(In1, In2).

$$CC0(Out) = \min\{CC0(In1) + CC0(In2), CC1(In1) + CC1(In2)\} + 1; \quad (4.13)$$

$$CC1(Out) = \min\{CC1(In1) + CC0(In2), CC0(In1) + CC1(In2)\} + 1; \quad (4.14)$$

$$OC(In1) = \min\{CC0(In2), CC1(In2)\} + OC(Out) + 1. \quad (4.15)$$

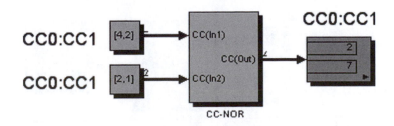

FIGURE 4.12
An example of CC0, CC1 calculations using the Simulink model of the NOR gate.

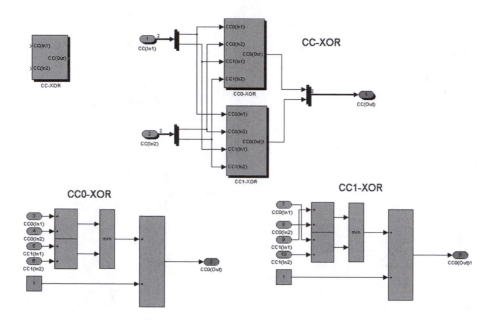

FIGURE 4.13
Simulink model of the XOR gate for the CC0, CC1 calculation at its output from the known CC0, CC1 values at the gate inputs.

The corresponding Simulink model for the CC0, CC1 calculations is shown in Figure 4.13, and the inner structures of the CC0-XOR and CC1-XOR units are constructed according to Equations 4.13 and 4.14.

Figure 4.14 gives a numerical example, which shows that if the logical zero, {CC0(In1) = 4} state, is set at the In1 input following four elementary operations, and the {CC1(In1) = 2} state is set following two elementary operations and at the same time, if the logical zero is set at the In2 input following two elementary operations, {CC0(In2) = 2}, then the gate output Out can be set to logical zero following four elementary operations, and to logical unity following five elementary operations (cf. Equations 4.13 and 4.14).

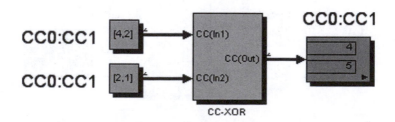

FIGURE 4.14
An example of CC0, CC1 calculations using the Simulink model of the XOR gate.

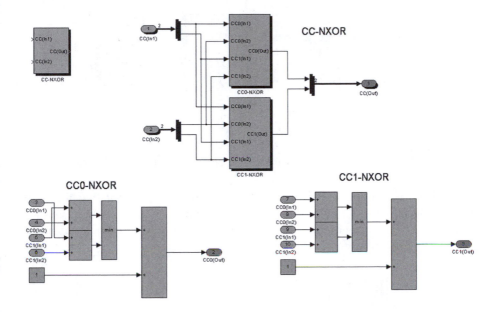

FIGURE 4.15
Simulink model of the NXOR gate for the CC0, CC1 calculation at its output from the known CC0, CC1 values at the gate inputs.

4.1.6 The NXOR Gate

Out = NXOR(In1, In2).

$$CC0(Out) = \min\{CC1(In1) + CC0(In2), CC0(In1) + CC1(In2)\} + 1; \quad (4.16)$$

$$CC1(Out) = \min\{CC0(In1) + CC0(In2), CC1(In1) + CC1(In2)\} + 1; \quad (4.17)$$

$$OC(In1) = \min\{CC0(In2), CC1(In2)\} + OC(Out) + 1. \quad (4.18)$$

The corresponding Simulink model for the CC0, CC1 calculations is shown in Figure 4.15, and the inner structures of the CC0-NXOR and CC1-NXOR units are constructed according to Equations 4.16 and 4.17. Figure 4.16 gives a numerical example, which shows that if the logical zero, {CC0(In1) = 4} state, is set at the In1 input following four elementary operations and the logical unity, {CC1(In1) = 2} state, is set following two elementary operations, and at the same time, the logical zero is set at the In2 input following one elementary operation, {CC0(In2) = 1}, then the gate output Out can be set to logical zero following five elementary operations and to logical unity, following four elementary operations (cf. Equations 4.16 and 4.17).

Simulink models of the XOR and NXOR gates for OC calculations are identical as follows from the comparison of the corresponding Equation 4.15 and Equation 4.18 expressions. The model is shown in Figure 4.17. The corresponding

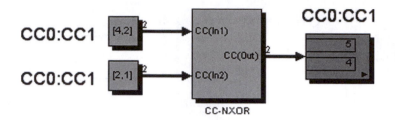

FIGURE 4.16
An example of CC0, CC1 calculations using the Simulink model of the NXOR gate.

numerical example is shown in Figure 4.18, and it is evident that the gate output logical state is transferred to one of its inputs by performing a certain number of elementary operations sufficient for setting its alternative input to the logical zero (the Equation 4.18 formula). To calculate the OC parameter

OC-XOR

OC-XOR

FIGURE 4.17
Simulink model of the XOR gate for OC calculations.

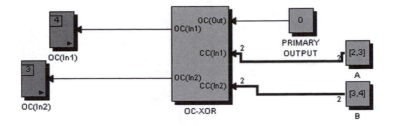

FIGURE 4.18
An example of OC calculations using the Simulink model of the XOR gate.

values for a gate input, we can use the CC0 parameter values for the rest of its inputs as well as the OC parameter value for the gate output.

4.1.7 The NOT Gate

Out = NOT(In).

$$CC0(Out) = CC1(In) + 1; \qquad (4.19)$$
$$CC1(Out) = CC0(In) + 1; \qquad (4.20)$$
$$OC(In) = OC(Out) + 1. \qquad (4.21)$$

The corresponding Simulink model for the CC0, CC1 calculations is shown in Figure 4.19. The inner structure of the CC0-NOT and CC1-NOT units corresponds to the Equation 4.19 and Equation 4.20 expressions.

Figure 4.20 gives a numerical example, which shows that if the logical unity is set at the In input following four elementary operations, {CC0(In) = 4}, and the logical zero is set at the In input following two elementary operations, {CC1(In) = 2}, then at the gate output Out can be set to logical zero, using three elementary operations and to the logical unity following five elementary operations (cf. Equations 4.19 and 4.20). The Simulink model for the NOT gate for the OC calculation is shown in Figure 4.21. The model is very simple and is constructed according to Equation 4.21.

We will now cite the example of the calculation of the indicated parameters for the nodes of the Figure 4.22 circuit. The calculation model for the CC0, CC1 parameters for each node of the Figure 4.22 circuit is shown in Figure 4.23. The input data for said calculations are given using the Constant Blocks (one block for each circuit input):

$$[CC0(1), CC1(1)] = [1,1];$$
$$[CC0(2), CC1(2)] = [1,1];$$
$$[CC0(3), CC1(3)] = [1,1].$$

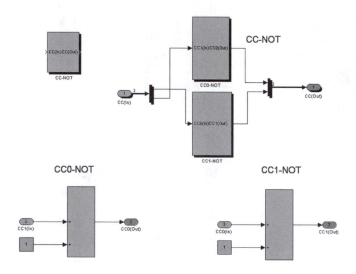

FIGURE 4.19
Simulink model of the NOT gate for the CC0, CC1 calculation at its output from the known CC0, CC1 values at the gate inputs.

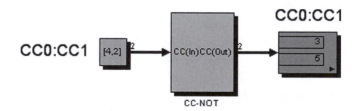

FIGURE 4.20
An example of CC0, CC1 calculations using the Simulink model of the NOT gate.

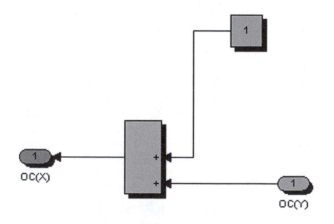

FIGURE 4.21
Simulink model of the NOT gate for OC calculations.

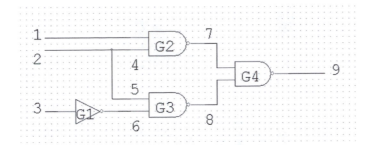

FIGURE 4.22
The circuit whose nodes need the calculation of the CC0, CC1 and OC parameters.

Seven Display Blocks are intended for mapping of the CC0, CC1 parameter values following all calculations:

Node 1: CC0:CC1_1 = [1,1]

Nodes 2, 4, and 5: CC0:CC1_2 = [1,1]

Node 3: CC0:CC1_3 = [1,1]

Node 6: CC0:CC1_6 = [2,2]

Node 7: CC0:CC1_7 = [3,2]

Node 8: CC0:CC1_8 = [4,2]

Node 9: CC0:CC1_9 = [5,4]

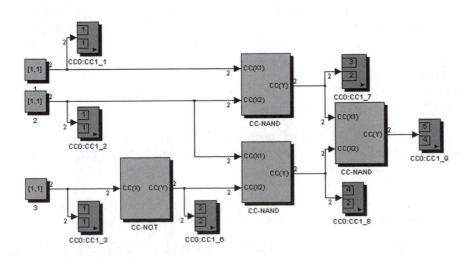

FIGURE 4.23
Simulink model of the Figure 4.22 circuit for the calculation of the CC0, CC1 parameters for each of its nodes.

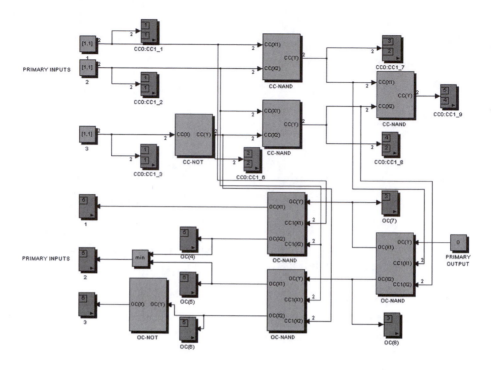

FIGURE 4.24
Simulink model of the Figure 4.22 circuit for the calculation of CC0, CC1 and OC parameters
for each of its nodes.

The circuit gates (three NAND gates and one NOT gate) are represented by
their Simulink models (CC_NAND and CC_NOT), discussed above, for the
calculation of the CC0, CC1 parameters from Equations 4.1, 4.2, 4.19, and 4.20.

Figure 4.24 displays the model for the CC0, CC1 and OC parameter
calculations in all nodes of the circuit being studied. The model's upper
portion (for the calculation of the CC0, CC1 parameters) was discussed
above. Now we will focus on the model's lower section. The input data for
the OC parameter calculations are:

1. The OC parameter value for the circuit-only output (for all primary
   outputs, the value equals zero by definition, because each circuit
   output can be observable at any moment of time) is set using the
   Constant Block.

2. The values of the CC0, CC1 parameters are introduced into the model's
   lower section from the upper one devised for their calculation.

Eight Display Blocks are devised for mapping the OC parameter calculation results in the circuit nodes (except node 9, the circuit output, where the
OC = 0 value is set):

Node 8: OC(8) = 3
Node 7: OC(7) = 3
Node 6: OC(6) = 5
Node 5: OC(5) = 6
Node 4: OC(4) = 5
Node 3: OC(3) = 6
Node 2: OC(2) = 5
Node 1: OC(1) = 5

The circuit gates (three NAND gates and one NOT gate) are represented by their Simulink models (OC_NAND and OC_NOT; discussed above) for the OC parameter calculations from Equations 4.3 and 4.21 (as already noted, the calculations flow moves from right to left — from primary outputs to primary inputs). The MinMax Block (having the minimum search directions) is placed on the circuit fanout stems (in our circuit, Figure 4.22 shows input [node] 2 connecting to internal nodes 4 and 5), as it is evident that the OC parameter value for the fanout stem input is minimal of the two OC parameter values for its outputs (branches).

When input stimuli on the circuit's primary inputs correspond to the random test data, CO = 0.5 for primary inputs (the probability of the logical unity occurrence at a circuit input equals the probability of the logical zero occurrence at a given input) since the circuit's primary outputs are always available for observation then OB = 1 for primary outputs. These are the boundary conditions for the calculation of CO and OB values in the circuit nodes. Calculation of CO values in circuit nodes is made from primary inputs to primary outputs, whereas the calculation of OB values goes in the opposite direction — from primary outputs to primary inputs. Both processes can take place in parallel, within the same model.

Let us consider the calculation formulas for CO and OB values of basic gates:

CO1 Calculation [1, 3]

$$\text{AND: } CO1(Out) = CO1(In1) \times CO1(In2) \tag{4.22}$$
$$\text{NAND: } CO1(Out) = 1 - CO1(In1) \times CO1(In2) \tag{4.23}$$
$$\text{OR: } CO1(Out) = 1 - (1 - CO1(In1)) \times (1 - CO1(In2)) \tag{4.24}$$
$$\text{NOR: } CO1(Out) = (1 - CO1(In1)) \times (1 - CO1(In2)) \tag{4.25}$$
$$\text{XOR: } CO1(Out) = 1 - (1 - CO1(In1)) \times (1 - CO1(In2)) \tag{4.26}$$
$$- CO1(In1) \times CO1(In2)$$

$$\text{NXOR: CO1(Out)} = (1 - \text{CO1(In1)}) \times (1 - \text{CO1(In2)}) \tag{4.27}$$

$$+ \text{CO1(In1)} \times \text{CO1(In2)}$$

CO0 Calculation [1, 3]

$$\text{AND: CO0(Out)} = 1 - (1 - \text{CO0(In1)}) \times (1 - \text{CO0(In2)}) \tag{4.28}$$

$$\text{NAND: CO0(Out)} = (1 - \text{CO0(In1)}) \times (1 - \text{CO0(In2)}) \tag{4.29}$$

$$\text{OR: CO0(Out)} = \text{CO0(In1)} \times \text{CO0(In2)} \tag{4.30}$$

$$\text{NOR: CO0(Out)} = 1 - \text{CO0(In1)} \times \text{CO0(In2)} \tag{4.31}$$

$$\text{XOR: CO0(Out)} = (1 - \text{CO0(In1)}) \times (1 - \text{CO0(In2)}) + \text{CO0(In1)} \times \text{CO0(In2)} \tag{4.32}$$

$$\text{NXOR: CO0(Out)} = 1 - (1 - \text{CO1(In1)}) \times (1 - \text{CO1(In2)}) - \text{CO1(In1)} \times \text{CO1(In2)} \tag{4.33}$$

OB Calculation [1, 3]

$$\text{AND: OB(In1)} = \text{OB(Out)} \times \text{CO1(In2)} \tag{4.34}$$

$$\text{NAND: OB(In1)} = \text{OB(Out)} \times \text{CO1(In2)} \tag{4.35}$$

$$\text{OR: OB(In1)} = \text{OB(Out)} \times (1 - \text{CO1(In2)}) \tag{4.36}$$

$$\text{NOR: OB(In1)} = \text{OB(Out)} \times (1 - \text{CO1(In2)}) \tag{4.37}$$

Now we construct the calculation model for the CO1 and OB parameters for the circuit in Figure 4.25. This model (Figure 4.26) obtains input data

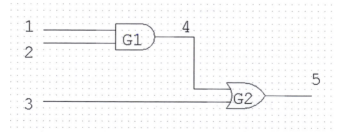

FIGURE 4.25
Circuit 1, whose nodes need the calculation of the CO1 and OB parameters.

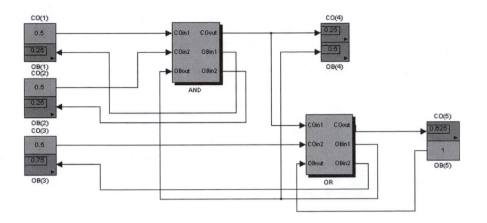

FIGURE 4.26
Simulink model of the Figure 4.25 circuit 1 for the calculation of the CO1 and OB parameters for each of its nodes.

from the Constant Blocks: CO1(1) = CO1(2) = CO1(3) = 0.5 (the CO1 parameter value for all circuit inputs by definition); OB(5) = 1 (the OB parameter value for all circuit outputs by definition).

Six Display Blocks are intended for calculation results mapping the CO1, OC parameters in the circuit nodes:

Node 1: OB(1) = 0.25 (circuit input)

Node 2: OB(2) = 0.25 (circuit input)

Node 3: OB(3) = 0.75 (circuit input)

Node 4: CO1(4) = 0.25, OB(4) = 0.5 (circuit internal node)

Node 5: CO1(5) = 0.625 (circuit output)

The circuit gates (the AND gate and OR gate) are represented by their Simulink models discussed above for the CO1, OB parameter calculations from Equations 4.22, 4.24, 4.34, and 4.36 (as already noted, the calculation's flow moves from right to left — from primary outputs to primary inputs for OB parameter — and from left to right — primary inputs to primary outputs — for CO1 parameter). The internal structures of the AND and OR units are shown in Figures 4.27 and 4.28, respectively.

We cite another example of the calculation model construction for CO1 and OB parameters for the circuit of Figure 4.29. This model (Figure 4.30) obtains input data from the Constant Blocks:
CO1(1) = CO1(2) = CO1(3) = CO1(4) = CO1(5) = 0.5 (the CO1 parameter value for all circuit inputs by definition);
OB(9) = OB(10) = 1 (the OB parameter value for all circuit outputs by definition).

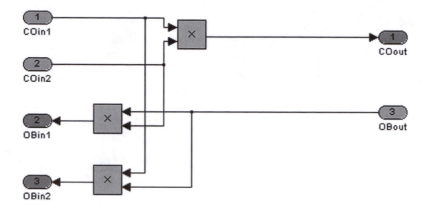

FIGURE 4.27
Simulink model of AND gate for the CO1 and OB calculation.

Thirteen Display Blocks are intended for calculation result mapping of the CO1, OC parameters in the circuit nodes:

Node 1: OB(1) = 0.1875 (circuit input)
Node 2: OB(2) = 0.0625 (circuit input)
Node 3: OB(3) = 0.0625 (circuit input)
Node 4: OB(4) = 0.5 (circuit input)

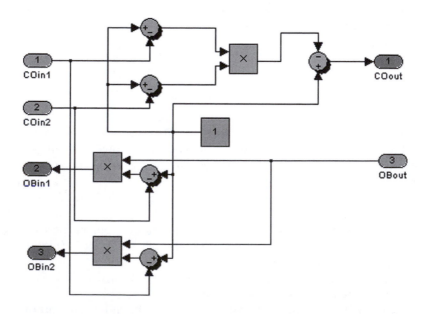

FIGURE 4.28
Simulink model of OR gate for the CO1 and OB calculation.

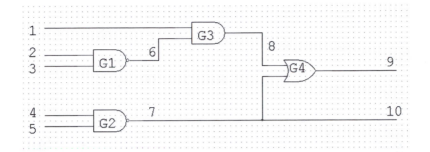

FIGURE 4.29
Circuit 2, whose nodes need the calculation of the CO1 and OB parameters.

Node 5: OB(5) = 0.5 (circuit input)
Node 6: CO1(6) = 0.75, OB(6) = 0.125 (circuit internal node)
Node 7: CO1(7) = 0.75, OB(7) = 1 (circuit internal node)
Node 8: CO1(8) = 0.375, OB(8) = 0.25 (circuit internal node)
Node 9: CO1(4) = 0.8438 (circuit output)
Node 10: CO1(5) = 0.75 (circuit output)

The circuit gates (the AND gate, two NAND gates, and the OR gate) are represented by their Simulink models discussed above for the CO1, OB

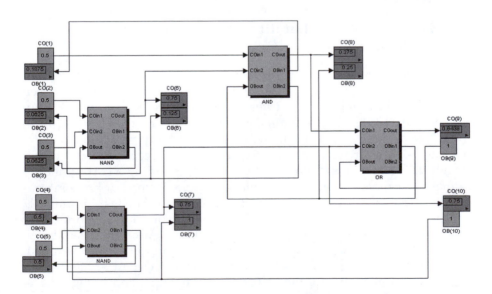

FIGURE 4.30
Simulink model of the Figure 4.25 circuit 2 for the calculation of the CO1 and OB parameters for each of its nodes.

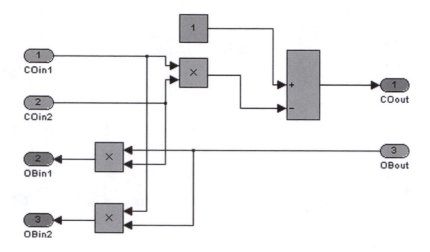

FIGURE 4.31
Simulink model of NAND gate for the CO1 and OB calculation.

parameter calculations from Equations 4.22, 4.23, 4.24, 4.35, and 4.36. The internal structure of the NAND unit is shown in Figure 4.31.

4.2 Sequential Controllability and Observability Analysis Models

We will now proceed to the assessment of testability parameters for sequential circuits. Figure 4.32(A) shows the sequential circuit composed of two FF1, FF2 flip-flops with corresponding feedbacks (from the FF1 flip-flop output to G4, G5 gate inputs, and from the FF2 flip-flop output to the G2 gate input). The feasible method of testability parameter calculation for such a circuit is its transformation into a combinational circuit (Figure 4.32(B)) by eliminating the flip-flops and adding a corresponding number of primary inputs and primary outputs instead of broken feedbacks (in this case, the first feedback is substituted for primary input PPI7 and primary output PPO7, whereas the second feedback is substituted for primary input PPI8 and primary output PPO8). After that the above-discussed models can be applied to the combinatorial circuit obtained. Such circuits for the Figure 4.32(B) transformed circuit are shown in Figure 4.33 (calculation of the CC0, CC1 parameters) and in Figure 4.34 (calculation of the OC parameter).

The simulation results obtained for the Figure 4.32(B) circuit using said models are the following:

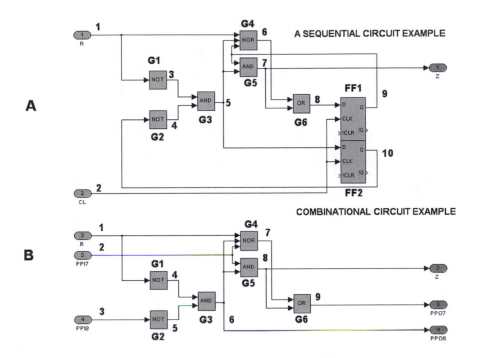

FIGURE 4.32
Sequential circuit and its transformation into a combinational circuit by elimination of flip-flops and addition of the appropriate number of primary inputs and primary outputs instead of the broken feedbacks.

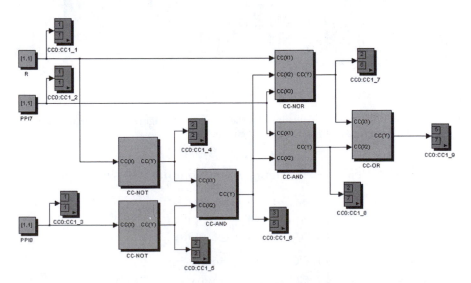

FIGURE 4.33
The model for CC0, CC1 parameter calculations for the nodes of the Figure 4.32(A) circuit, pretransformed into the Figure 4.32(B) combinational circuit.

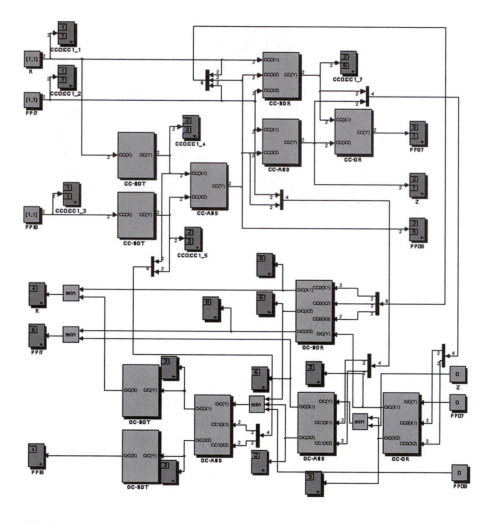

FIGURE 4.34

The model for OC parameter calculations for the nodes of the Figure 4.32(A) circuit, pretransformed into the Figure 4.32(B) combinational circuit.

The CC0, CC1 parameters (Figure 4.33) (input data for simulation):

$$CC0(R) = 1, CC1(R) = 1$$
$$CC0(PPI7) = 1, CC1(PPI7) = 1$$
$$CC0(PPI8) = 1, CC1(PPI8) = 1)$$

Node 1: $CC0(1) = 1, CC1(1) = 1$ (primary input R)

Node 2: $CC0(2) = 1, CC1(2) = 1$ (primary input PPI7)

Node 3: $CC0(3) = 1, CC1(3) = 1$ (primary input PPI8)

Node 4: $CC0(4) = 2, CC1(4) = 2$ (circuit internal node; G1 gate output)

Node 5: CC0(5) = 2, CC1(5) = 2 (circuit internal node; G2 gate output)
Node 6: CC0(6) = 3, CC1(6) = 5 (primary output PPO8)
Node 7: CC0(7) = 2, CC1(7) = 6 (circuit internal node; G4 gate output)
Node 8: CC0(8) = 2, CC1(8) = 7 (primary output Z)
Node 9: CC0(9) = 5, CC1(9) = 7 (primary output PPO7)

The OC parameter (Figure 4.34) (input data for simulation):

$$OC(Z) = 0$$
$$OC(PPO7) = 0$$
$$OC(PPO8) = 0)$$

Node 1: OC(1) = 4 (primary input R)
Node 2: OC(2) = 6 (primary input PPI7)
Node 3: OC(3) = 4 (primary input PPI8)
Node 4: OC(4) = 3 (circuit internal node; G1 gate output)
Node 5: OC(5) = 3 (circuit internal node; G2 gate output)
Node 6: OC(6) = 0 (primary output PPO8)
Node 7: OC(7) = 3 (circuit internal node; G4 gate output)
Node 8: OC(8) = 0 (primary output Z)
Node 9: OC(9) = 0 (primary output PPO7)

We will now proceed to the calculation of the sequential controllability (SC) and sequential observability (SO) parameters for the sequential circuit nodes (sequential SCOAP measures). In this process we use the formulas for the calculation of CC0,CC1; SC0,SC1; and CO,SO parameters from [1] for synchronously resettable negative-edge-triggered D-type flip-flop (DFF):

$$CC1(Q) = CC1(D) + CC1(C) + CC0(C) + CC0(RESET); \qquad (4.38)$$
$$SC1(Q) = SC1(D) + SC1(C) + SC0(C) + SC0(RESET) + 1; \qquad (4.39)$$
$$CC0(Q) = \min [CC1(RESET) + CC1(C) + CC0(C), CC0(D)$$
$$+ CC1(C) + CC0(C)]; \qquad (4.40)$$
$$SC0(Q) = \min [SC1(RESET) + SC1(C) + SC0(C), SC0(D)$$
$$+ SC1(C) + SC0(C)] + 1; \qquad (4.41)$$
$$CO(D) = CO(Q) + CC1(C) + CC0(C) + CC0(RESET); \qquad (4.42)$$
$$SO(D) = SO(Q) + SC1(C) + SC0(C) + SC0(RESET) + 1; \qquad (4.43)$$
$$CO(RESET) = CO(Q) + CC1(Q) + CC1(RESET) + CC1(C) + CC0(C); \qquad (4.44)$$
$$SO(RESET) = SO(Q) + SC1(Q) + SC1(RESET) + SC1(C) + SC0(C) + 1; \qquad (4.45)$$

$$CO(C) = \min [CO(Q) + CC1(Q) + CC0(D) + CC1(C) + CC0(C), CO(Q)$$
$$+ CC1(Q) + CC1(RESET) + CC1(C) + CC0(C), CO(Q)$$
$$+ CC0(Q) + CC0(RESET) + CC1(D) + CC1(C) + CC0(C)]; \quad (4.46)$$
$$SO(C) = \min [SO(Q) + SC1(Q) + SC0(D) + SC1(C) + SC0(C), SO(Q)$$
$$+ SC1(Q) + SC1(RESET) + SC1(C) + SC0(C), SO(Q) + SC0(Q)$$
$$+ SC0(RESET) + SC1(D) + SC1(C) + SC0(C)] + 1. \quad (4.47)$$

In addition to the aforesaid expressions, we can apply the value calculation algorithm of combinational and sequential measures from [1] for the model construction:

1. The following initial values should be given for all primary inputs — PIs of the I circuit:

$$CC0(I) = CC1(I) = 1; \quad (4.48)$$
$$SC0(I) = SC1(I) = 1. \quad (4.49)$$

2. For any other N node of the circuit, the following initial values are established:

$$CC0(N) = CC1(N) = \infty; \quad (4.50)$$
$$SC0(N) = SC1(N) = \infty. \quad (4.51)$$

3. Similar parameter values for the gate and flip-flop outputs are calculated in the direction from the circuit primary inputs to primary outputs, using the equations for CC0, CC1, SC0, and SC1 at the gate and flip-flop inputs. Such calculations can be iterative if the circuit contains feedback loops and can proceed until the CC0, CC1, SC0, and SC1 stable values are obtained.

4. Then the following initial values are established for all primary outputs — POs of the U circuit:

$$CO(U) = SO(U) = 0. \quad (4.52)$$

5. For any other N node of the circuit, the following initial values are established:

$$CO(N) = SO(N) = \infty. \quad (4.53)$$

6. The CO and SO parameter values for the gate and flip-flop outputs are calculated in the direction from the circuit primary outputs to primary inputs, using the equations for CO and SO at the gate and flip-flop inputs and outputs, as well as previously calculated CC0, CC1, SC0, and SC1. For fanout stems Z with Z1,..., ZN branches, the calculations are based on

$$SO(Z) = \min(SO(Z1),..., SO(ZN)); \quad (4.54)$$
$$CO(Z) = \min(CO(Z1),..., CO(ZN)). \quad (4.55)$$

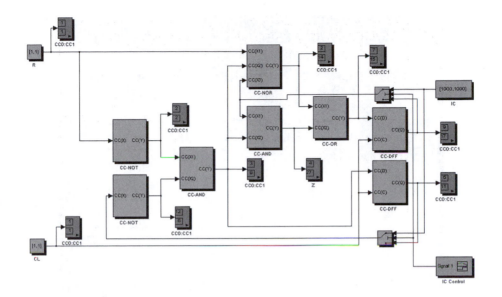

FIGURE 4.35
The model for the CC0, CC1 parameter calculations for the nodes of the Figure 4.32(A) sequential circuit.

7. If the calculation results show that

For a certain node CC0(SC0) = ∞, it means that node is 0–uncontrollable

For a certain node CC1(SC1) = ∞, it means that node is 1–uncontrollable)

For a certain node CO = ∞ or SO = ∞, it means that node is unobservable.

The model shown in (Figure 4.35) can be plotted based on the above formulas and algorithm. The model obtains input data from the green Constant Blocks:

$$CC0(R) = 1, CC1(R) = 1;$$
$$CC0(CL) = 1, CC1(CL) = 1$$

(the CC0, CC1 parameter values for all circuit inputs are taken from Equation 4.48).

The circuit components (two AND gates, two NOT gates, a NOR gate, an OR gate, and two DFF — cyan) are represented by their Simulink models (CC_AND, CC_NOT, _NOR, CC_OR, CC_DFF) for the calculation of the CC0, CC1 parameters in sequential circuits from Equations 4.1, 4.2, 4.19, 4.20, 4.10, 4.11, 4.7, 4.8, 4.38, and 4.40, respectively.

The Constant Block labeled IC and the Signal Builder Block labeled IC Control, as well as two Switch Blocks, are devised for starting conditions prior to simulation according to Equation 4.50 (instead of ∞ abstract values,

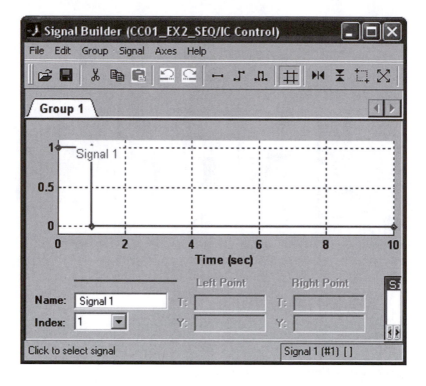

FIGURE 4.36
The starting condition entering signal generated by the Signal Builder Block.

certain large values are entered, in this case 1000). The signal whose time diagram is shown in Figure 4.36 is the signal for entering the initial conditions.

Ten Display Blocks are intended for mapping of CC0, CC1 parameter values following all calculations:

Node 1: CC0:CC1_1 = [1,1] (primary input R)
Node 2: CC0:CC1_2 = [1,1] (primary input CL)
Node 3: CC0:CC1_3 = [2,2] (circuit internal node; NOT G2 gate output)
Node 4: CC0:CC1_4 = [12,6] (circuit internal node; NOT G2 gate output)
Node 5: CC0:CC1_5 = [3,9] (circuit internal node; AND G3 gate output)
Node 6: CC0:CC1_6 = [2,14] (circuit internal node; NOR G4 gate output)
Node 7: CC0:CC1_7 = [4,27] (primary output Z)
Node 8: CC0:CC1_8 = [7,15] (circuit internal node; OR G6 gate output)
Node 9: CC0:CC1_9 = [9,17] (circuit internal node; FF1 flip-flop output)
Node 10: CC0:CC1_10 = [5,11] (circuit internal node; FF2 flip-flop output)

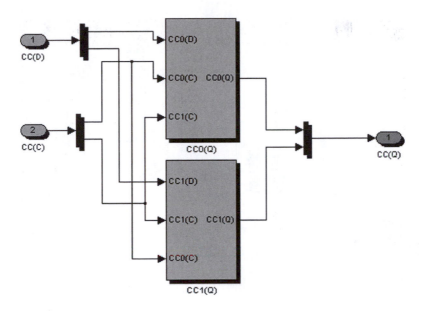

FIGURE 4.37
The CC_DFF unit's internal structure.

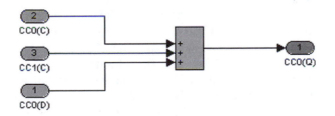

FIGURE 4.38
The CC0(Q) unit's internal structure.

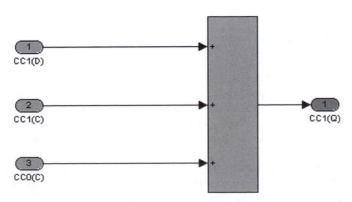

FIGURE 4.39
The CC1(Q) unit's internal structure.

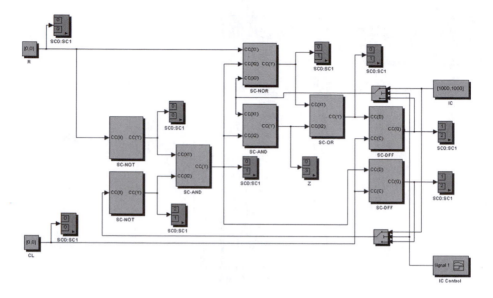

FIGURE 4.40

The model for the SC0, SC1 parameter calculations for the nodes of the Figure 4.32a sequential circuit.

The model unit structures of CC_AND, CC_NOT, _NOR, and CC_OR have already been discussed. The CC_DFF unit structure is shown in Figures 4.37 to 4.39. It is constructed according to Equations 4.38 and 4.40.

Figure 4.40 shows the model for the SC0, SC1 parameter calculations for the sequential circuit in Figure 4.32(A). The model is nearly identical to the model just discussed. It should be noted that the inner structures of the SC_AND, SC_NOT, S_NOR, SC_OR, and SC_DFF units are plotted from the expressions lacking right-hand additional unity (as compared to Equations 4.1 and 4.2, 4.19 and 4.20, 4.10 and 4.11, 4.7 and 4.8, and 4.39 and 4.40).

References

1. Bushnell M.L., Agrawal V.D. Essentials of Electronic Testing for Digital, Memory & Mixed-Signal VLSI Circuits. Kluwer Academic Publishers, Dordrecht, The Netherlands, 2004.
2. Navabi Z. VHDL: Analysis and Modeling of Digital Systems. McGraw-Hill, Singapore, 1998.
3. Smith M.J.S. Application-Specific Integrated Circuits. Addison-Wesley, Reading, MA, 1997.

5

The Automatic Test Pattern Generation (ATPG) Process

5.1 ATPG Fundamentals

The D-calculus is used for test (integrity of test vectors) development for digital circuits. The treatment enables parallel description of fault-free circuit (good circuit) and faulty circuit (bad circuit) behavior. In this process the composite logic value D (meaning *for detect*) is a logical unity for a fault-free circuit and a logical zero for the circuit containing one or more faults. Instead of D, 1/0 could have been used [1–2].

The formal definition of the D logic value obtained at a circuit output following D occurrence at one or at several of its inputs is as follows.

$$D = \begin{cases} 1 - \text{for fault-free circuit;} \\ 0 - \text{for faulty circuit.} \end{cases}$$

Similarly, D inverse value can be introduced:

$$\bar{D} = \begin{cases} 1 - \text{for faulty circuit;} \\ 0 - \text{for fault-free circuit.} \end{cases}$$

Let us examine an OR gate with two inputs (Figure 5.1). Figure 5.1(A) depicts a fault-free specimen of the gate that can be described by a simple truth table (Table 5.1). Figure 5.1(B) depicts a specimen of the gate that contains the *constant zero* fault at the B − s@0 (stack-at-0) input, the fault that is also described by the corresponding truth table (Table 5.2). Combining these two descriptions, we obtain the unified OR logical gate (Figure 5.1(C)), which looks like the following logical cube (Table 5.3). The description fits both the fault-free OR gate and the gate with the constant zero fault at the B input.

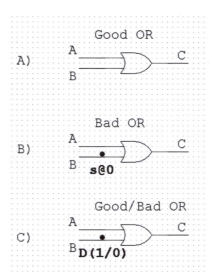

FIGURE 5.1
Description of an OR logical gate using "the cubes."

TABLE 5.1

Truth Table for Good OR (Figure 5.1(A))

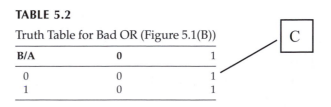

B/A	0	1
0	0	1
1	1	1

TABLE 5.2

Truth Table for Bad OR (Figure 5.1(B))

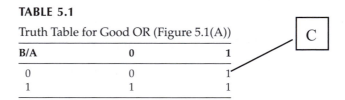

B/A	0	1
0	0	1
1	0	1

TABLE 5.3

Truth Table for Good/Bad OR (Figure 5.1(C))

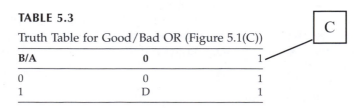

B/A	0	1
0	0	1
1	D	1

TABLE 5.4

The Primitive Cube for the AND Gate

A	B	C
0	X	0
X	0	0
1	1	1

In test development, various logical cubes are used [1–3]:

1. **Primitive cube:** The operation of fault-free gates
2. **Failure cube:** The operation of faulty gates
3. **Propagate cube:** The passage of D, $\overline{D}$ logic values from the gate inputs to their outputs (for each such cube a dual cube where D is substituted for $\overline{D}$ can be composed immediately)
4. **Nonpropagate cube:** Conditions when the passage of D, $\overline{D}$ logic values from gate inputs to their outputs is impracticable

Following are the four cubes for major gate types:

1. The AND gate
   Primitive cube (Table 5.4)
   Failure cube (Table 5.5)
   Propagate cube (Table 5.6)
   Nonpropagate cube (Table 5.7)
2. The OR gate
   Primitive cube (Table 5.8)
   Failure cube (Table 5.9)
   Propagate cube (Table 5.10)
   Nonpropagate cube (Table 5.11)
3. The NAND gate
   Primitive cube (Table 5.12)

TABLE 5.5

The Failure Cube for the AND Gate

A	B	C	
0	0	$\overline{D}$	C s@1
0	1	$\overline{D}$	A s@1, C s@1
1	0	$\overline{D}$	B s@1, C s@1
1	1	D	A s@0, B s@0, C s@0

TABLE 5.6

The Propagate Cube for the AND Gate

A	B	C
1	D	D
D	1	D
D	D	D

TABLE 5.7

The Nonpropagate Cube for the AND Gate

A	B	C
0	D	0
D	0	0
D	$\overline{D}$	0

TABLE 5.8

The Primitive Cube for the OR Gate

A	B	C
1	X	1
X	1	1
0	0	0

TABLE 5.9

The Failure Cube for the OR Gate

A	B	C	
1	0	D	A s@0, C s@0
0	1	D	B s@0, C s@0
0	0	$\overline{D}$	A s@0, B s@0, C s@1
1	1	D	C s@1

TABLE 5.10

The Propagate Cube for the OR Gate

A	B	C
0	D	D
D	0	D
D	D	D

TABLE 5.11

The Nonpropagate Cube for the OR Gate

A	B	C
1	D	1
D	1	1
D	$\overline{D}$	1

TABLE 5.12

The Primitive Cube for the NAND Gate

A	B	C
0	X	1
X	0	1
1	1	0

Failure cube (Table 5.13)
Propagate cube (Table 5.14)
Nonpropagate cube (Table 5.15)
4. The NOR gate
Primitive cube (Table 5.16)
Failure cube (Table 5.17)
Propagate cube (Table 5.18)
Nonpropagate cube (Table 5.19)
5. The NOT gate
Primitive cube (Table 5.20)
Failure cube (Table 5.21)
Propagate cube (Table 5.22)
6. The connector
Primitive cube (Table 5.23)
Failure cube (Table 5.24)
Propagate cube (Table 5.25)
7. The XOR (Exclusive - OR)
Primitive cube (Table 5.26)
Failure cube (Table 5.27)
Propagate cube (Table 5.28)
Nonpropagate cube (Table 5.29)

TABLE 5.13

The Failure Cube for the NAND Gate

A	B	C	
0	D	1	B s@1, C s@0
D	0	1	A s@1, C s@0
1	1	$\overline{D}$	A s@0, B s@0, C s@1
0	0	D	C s@0

TABLE 5.14

The Propagate Cube for the NAND Gate

A	B	C
1	D	$\overline{D}$
D	1	$\overline{D}$
D	D	$\overline{D}$

TABLE 5.15

The Nonpropagate Cube for the NAND Gate

A	B	C
0	D	1
D	0	1
D	$\overline{D}$	1

TABLE 5.16

The Primitive Cube for the NOR Gate

A	B	C
1	X	0
X	1	0
0	0	1

TABLE 5.17

The Failure Cube for the NOR Gate

A	B	C	
0	1	$\overline{D}$	B s@0, C s@1
1	0	$\overline{D}$	A s@0, C s@1
0	0	D	A s@1, B s@1, C s@0
1	1	$\overline{D}$	C s@1

TABLE 5.18

The Propagate Cube for the NOR Gate

A	B	C
1	X	0
X	1	0
0	0	1

TABLE 5.19

The Nonpropagate Cube for the NOR Gate

A	B	C
0	D	$\overline{D}$
D	0	$\overline{D}$
D	D	$\overline{D}$

TABLE 5.20

The Primitive Cube for the NOT Gate

A	B
1	0
0	1

TABLE 5.21

The Failure Cube for the NOT Gate

A	B	
1	$\overline{D}$	A s@0, B s@1
0	D	A s@1, B s@0

TABLE 5.22

The Propagate Cube for the NOT Gate

A	B
D	$\overline{D}$

TABLE 5.23

The Primitive Cube for the Connector

A	B	C
1	1	1
0	0	0

TABLE 5.24

The Failure Cube for the Connector

A	B	C	
0	$\overline{D}$	$\overline{D}$	A s@0
1	D	D	A s@1

TABLE 5.25

The Propagate Cube for the Connector

A	B	C
D	D	D

TABLE 5.26

The Primitive Cube for the XOR Gate

A	B	C
0	0	0
0	1	1
1	0	1
1	1	0

All these cubes can be generalized using unified cubes (one cube for each gate) [2]; the AND gate (Table 5.30), the OR gate (Table 5.31), the NAND gate (Table 5.32), the NOR gate (Table 5.33), the XOR gate (Table 5.34), and the CONNECT connector (Table 5.35). The representation is the base for our gate models in this chapter. We apply the Look-Up Table (two-dimensional [2-D]) Block of the Simulink® (Figure 5.2) system.

The block contains a 2-D two-input (indices) matrix. If we set the appropriate values of two operands (corresponding to the chosen line and column), then at the block output we have the value found in the cell at the intersection of said line and column. The introduction of matrix component values to the block parameters window is shown in Figure 5.3.

In this process the codes shown in Table 5.36 are used for the cubes' logical values. The gate models and test bench for their trials are given in Figure 5.4. The cubes can be designed for entire circuits (circuit cubes) as well. To do this, all circuit nodes should be numbered in the sequence shown in Figure 5.5. Table 5.37 cites several examples of circuit cubes for the circuit in Figure 5.5.

For our cube operation we will use a single mathematical operation called the cube intersection: $\cap$. The operation is the basis for test vector findings. Take, for example, two cubes, $A = a_1 \, a_2 \, \dots \, a_i \, \dots \, a_n$ and $B = b_1 \, b_2 \, \dots \, b_i \, \dots \, b_n$, that describe the conditions of the selected n nodes of the same circuit. The cube intersection can be executed on these two cubes according to Table 5.38, where ξ denotes the conflict of an attempt to assign logical value to a node. For instance, if cube A states 1 ($a_i = 1$) for a certain node, and cube B defines 0 ($b_i = 0$) for the same node, then ξ appears in the corresponding line–column intersection cell and witnesses the impossibility of obtaining the cube intersection results for a given circuit node. The cube intersection technique is applicable to all cube types. We do not need to describe the states of all circuit nodes in one cube and can restrict ourselves to the description of

TABLE 5.27

The Failure Cube for the XOR Gate

A	B	C	
0	0	$\overline{D}$	A s@1, B s@1, C s@0
0	1	D	B s@0, C s@1
1	0	D	A s@0, C s@0
1	1	$\overline{D}$	C s@1

TABLE 5.28

The Propagate Cube for the XOR Gate

A	B	C
D	0	D
0	D	D
D	1	$\overline{D}$
1	D	$\overline{D}$

TABLE 5.29

The Nonpropagate Cube for the XOR Gate

A	B	C
D	D	0
D	$\overline{D}$	1
$\overline{D}$	D	1
$\overline{D}$	$\overline{D}$	0

TABLE 5.30

Generalized Cube for the AND Gate

A\B	0	1	D	$\overline{D}$	X
0	0	0	0	0	0
1	0	1	D	$\overline{D}$	X
D	0	D	D	0	X
$\overline{D}$	0	$\overline{D}$	0	$\overline{D}$	X
X	0	X	X	X	X

TABLE 5.31

Generalized Cube for the OR Gate

A\B	0	1	D	$\overline{D}$	X
0	0	1	D	$\overline{D}$	X
1	1	1	1	1	1
D	D	1	D	1	X
$\overline{D}$	$\overline{D}$	1	1	$\overline{D}$	X
X	X	1	X	X	X

TABLE 5.32

Generalized Cube for the NAND Gate

A\B	0	1	D	$\overline{D}$	X
0	1	1	1	1	1
1	1	0	$\overline{D}$	D	X
D	1	$\overline{D}$	D	1	X
$\overline{D}$	1	D	1	D	X
X	1	X	X	X	X

TABLE 5.33

Generalized Cube for the NOR Gate

A\B	0	1	D	$\overline{D}$	X
0	1	0	$\overline{D}$	D	X
1	0	0	0	0	0
D	$\overline{D}$	0	$\overline{D}$	0	X
$\overline{D}$	D	0	0	D	X
X	X	0	X	X	X

TABLE 5.34

Generalized Cube for the XOR Gate

A\B	0	1	D	$\overline{D}$	X
0	0	1	D	$\overline{D}$	X
1	1	0	$\overline{D}$	D	X
D	D	$\overline{D}$	0	1	X
$\overline{D}$	$\overline{D}$	D	1	0	X
X	X	X	X	X	X

TABLE 5.35

Generalized Cube for the Connector

A	B	C
0	0	0
1	1	1
D	D	D
$\overline{D}$	$\overline{D}$	$\overline{D}$
X	X	X

certain selected nodes, which brings us to the shorthand cubic notation. For instance, let the circuit contain 14 nodes so that data on their logical states exists only for three of them. The complete cube

$$X_1\ X_2 1_3\ X_4\ X_5\ X_6\ X_7 D_8\ D_9\ X_{10}\ X_{11}\ X_{12}\ X_{13}\ X_{14}$$

formally contains the logical values for all circuit nodes, but the indeterminacy value X gives no positive information, making the shorthand notation more expedient:

$$1_3\ D_8\ D_9.$$

We can cite two instances of the cube intersection results:

$$0\ 1\ X\ D\ X \cap 0\ X\ 1\ D\ \overline{D} = 0\ 1\ 1\ D\ \ \overline{D};$$
$$0\ 1\ X\ D\ X \cap 0\ 0\ X\ 1\ \overline{D} = 0\ \xi\ X\ \xi\ D$$

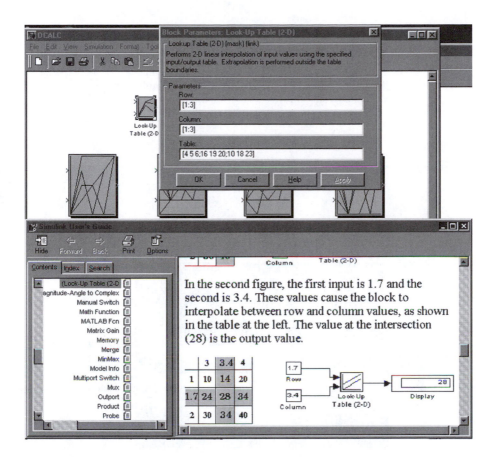

FIGURE 5.2
Simulink Look-Up Table (2-D) Block.

Thus, in the second instance we could not obtain the result, as the states of two nodes (2 and 4) differ for the two cubes that represent the intersection operands [3].

5.2 Combinational Circuit ATPG (Current-Based ATPG Algorithms for Combinational Circuits)

5.2.1 The D-Algorithm Model

We will now look at two examples of test vector findings using the D algorithm for the preset faults in two combinational circuits [3].

TABLE 5.36

Corresponding Codes for the Cube Logical Values

Cube Logical Value	Corresponding Code for the Operation in the Simulink Environment
0	0
1	1
D	2
$\overline{D}$	3
X	4

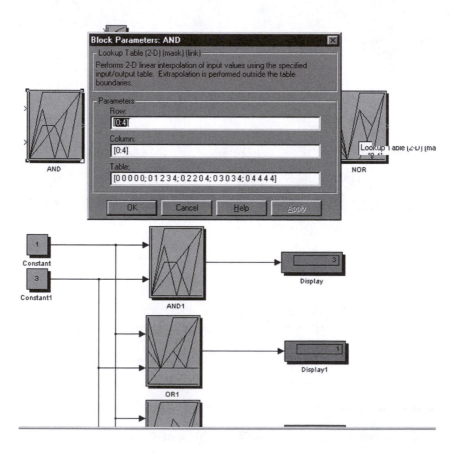

FIGURE 5.3
Introduction of the cubes' logical values in the parameter window of Simulink Look-Up Table (2-D) Block.

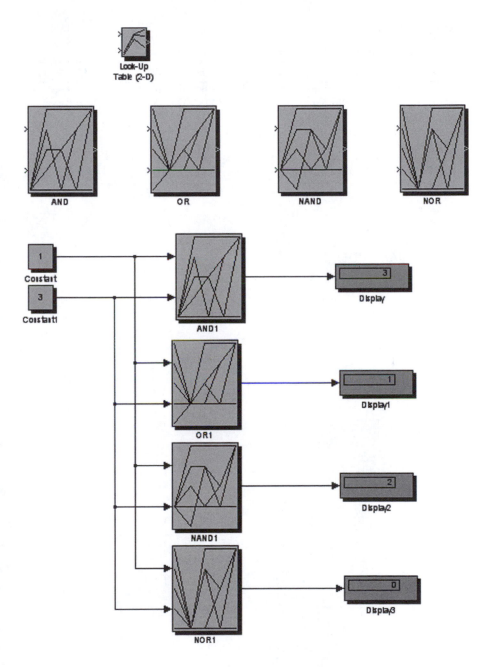

FIGURE 5.4
The gate models.

TABLE 5.37

Circuit Cubes for the Circuit in Figure 5.5

Circuit node No.	1	2	3	4	5	6	7	8	9	10	11	12	13	14	15
Cube for FFC	1	0	0	1	1	0	0	0	0	0	1	1	1	0	1
Cube for the 7 s@0) fault circuit	0	0	1	1	1	0	D	D	D	D	1	1	1	D	D
Failure cube for gate G3	X	X	X	X	X	0	X	D	X	D	X	X	X	X	X
Propagate cube for gate G5	X	X	X	X	X	X	X	X	X	D	X	0	X	D	X

TABLE 5.38

Cube Intersection Operation Definition

b_i / a_i	0	1	X	D	$\overline{D}$
0	0	ξ	0	ξ	ξ
1	ξ	1	1	ξ	ξ
X	0	1	X	D	$\overline{D}$
D	ξ	ξ	D	D	ξ
$\overline{D}$	ξ	ξ	$\overline{D}$	ξ	$\overline{D}$

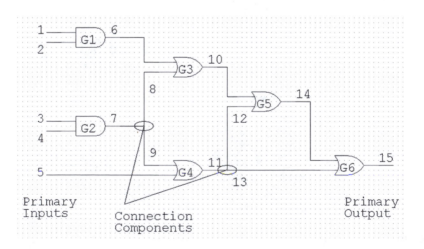

FIGURE 5.5
D-algorithm circuit notation.

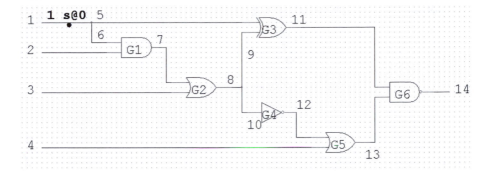

FIGURE 5.6
The circuit with a constant 0-type fault in node 1 (1 s@0).

Example 1

We must find the test vector for detecting the constant-0 in a one-node-type (1 s@0) fault in the Figure 5.6 circuit. The test vector findings occur as follows:

1. The Failure Cube is $1_1 \, D_5 \, D_6$.
2. Propagate through the XOR gate (G3):

$$1_1 \, D_5 \, D_6 \cap D_5 \, 1_9 \, \overline{D}_{11} = 1_1 \, D_5 \, D_6 \, 1_9 \, \overline{D}_{11}.$$

3. Propagate through the final NOR gate (G6):

$$1_1 \, D_5 \, D_6 \, 1_9 \, \overline{D}_{11} \cap \overline{D}_{11} \, 0_{13} \, D_{14} = 1_1 \, D_5 \, D_6 \, 1_9 \, \overline{D}_{11} 0_{13} \, D_{14}.$$

4. Back through the OR gate (G5):

$$1_1 \, D_5 \, D_6 \, 1_9 \, \overline{D}_{11} 0_{13} \, D_{14} \cap 0_4 \, 0_{12} \, 0_{13} = 1_1 \, 0_4 \, D_5 \, D_6 \, 1_9 \, \overline{D}_{11} \, 0_{12} \, 0_{13} \, D_{14}.$$

5. Back through the inverter (G4):

$$1_1 \, 0_4 \, D_5 \, D_6 \, 1_9 \, \overline{D}_{11} \, 0_{12} \, 0_{13} \, D_{14} \cap 1_{10} \, 0_{12} = 1_1 \, 0_4 \, D_5 \, D_6 \, 1_9 \, 1_{10} \, \overline{D}_{11} \, 0_{12} \, 0_{13} \, D_{14}.$$

6. Back through the connector at nodes 8, 9, and 10:

$$1_1 \, 0_4 \, D_5 \, D_6 \, 1_9 \, 1_{10} \, \overline{D}_{11} \, 0_{12} \, 0_{13} \, D_{14} \cap 1_8 \, 1_9 \, 1_{10}$$
$$= 1_1 \, 0_4 \, D_5 \, D_6 \, 1_8 \, 1_9 \, 1_{10} \, \overline{D}_{11} \, 0_{12} \, 0_{13} \, D_{14}.$$

7. Back through the OR gate (G2):

$$1_1 \, 0_4 \, D_5 \, D_6 \, 1_8 \, 1_9 \, 1_{10} \, \overline{D}_{11} \, 0_{12} \, 0_{13} \, D_{14} \cap 1_3 \, X_7 \, 1_8$$
$$= 1_1 1_3 \, 0_4 \, D_5 \, D_6 \, X_7 \, 1_8 \, 1_9 \, 1_{10} \, \overline{D}_{11} \, 0_{12} \, 0_{13} \, D_{14}.$$

8. Back through the AND gate (G1):

$$1_1 1_3 0_4 D_5 D_6 X_7 1_8 1_9 1_{10} \overline{D}_{11} 0_{12} 0_{13} D_{14} \cap 0_2 D_6 0_7$$
$$= 1_1 0_2 1_3 0_4 D_5 D_6 X_7 1_8 1_9 1_{10} \overline{D}_{11} 0_{12} 0_{13} D_{14}.$$

Application of the input vector (test vector) is $1_1 0_2 1_3 0_4$:

$$\text{Output} = 1 \text{ for a fault-free circuit;}$$
$$\text{Output} = 0 \text{ for a circuit with fault 1 s@0.}$$

Example 2

We must find the test vector for detecting the constant-1 in six-node-type (6 s@1) fault in the Figure 5.7 circuit:

1. The Failure Cube is $1_3 \overline{D}_6$.
2. Propagate through the NAND gate (G3):

$$1_3 \overline{D}_6 \cap 1_5 \overline{D}_6 D_8 = 1_3 1_5 \overline{D}_6 D_8.$$

3. Propagate through the NAND gate (G4):

$$1_3 1_5 \overline{D}_6 D_8 \cap 1_7 D_8 \overline{D}_9 = 1_3 1_5 \overline{D}_6 1_7 D_8 \overline{D}_9.$$

4. Back through the NAND gate (G2):

$$1_3 1_5 \overline{D}_6 1_7 D_8 \overline{D}_9 \cap 0_1 1_4 1_7 = 0_1 1_3 1_4 1_5 \overline{D}_6 1_7 D_8 \overline{D}_9.$$

5. Back through the connector at nodes 2, 4, and 5:

$$0_1 1_3 1_4 1_5 \overline{D}_6 1_7 D_8 \overline{D}_9 \cap 1_2 1_4 1_5 = 0_1 1_2 1_3 1_4 1_5 \overline{D}_6 1_7 D_8 \overline{D}_9.$$

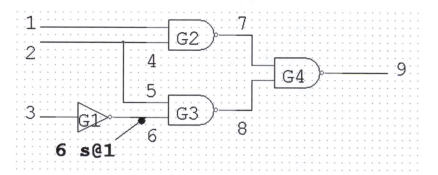

FIGURE 5.7
The circuit with a constant one-type fault in node 6 (6 s@1).

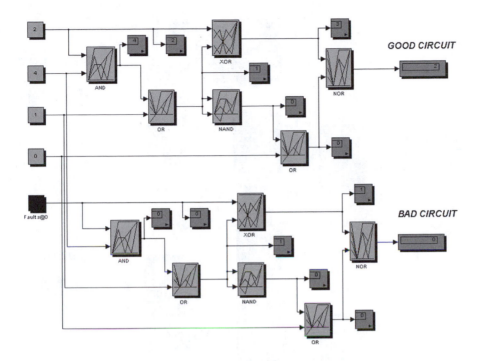

FIGURE 5.8
Simulink model of the D algorithm application in the Figure 5.6 circuit.

Application of the input vector (test vector) is $0_1 \, 1_2 \, 1_3$:

Output = 0 for a fault-free circuit;

Output = 1 for a circuit with fault 6 s@1.

We will now construct the Simulink models for the D algorithm application the two previous examples. The Simulink model for Example 1 is given in Figure 5.8. The D, X, 1, 0 vector at the circuit inputs (nodes 1, 2, 3, 4) represents the model input data. It is given by four Constant Blocks. The model is composed of two parts: a submodel for the fault-free circuit and a submodel for the faulty circuit. The Constant Block, representing the 0 constant, is the model of the 1 s@0 fault. The circuit gates in the Simulink model are represented by Look-Up Table (2-D) Blocks with corresponding AND, OR, XOR, NAND, and NOR names that contain generalized cubes of these gates. Seven Display Blocks for each submodel display the process of fault message propagation (D-frontier) from primary input 1 to primary output 14. It can be seen that in a good circuit the fault message arrives at the circuit output without distortions. Therefore, by setting the circuit primary input 1 to logical unity condition, we have the same condition at primary output 14 (for D = 1 fault-free circuit). If the circuit is faulty (primary input 1 is fixed at logical zero condition), we have the same logical zero value (for D = 0 faulty circuit) at primary output 14. The Simulink model for Example 2 given in Figure 5.9 is built similarly.

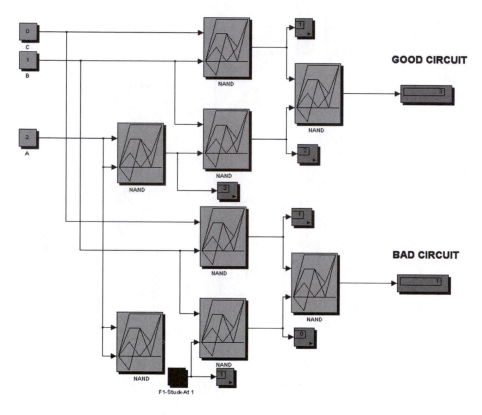

FIGURE 5.9
Simulink model of the D algorithm application in the Figure 5.7 circuit.

5.2.2 The PODEM-Algorithm Model

For many simple circuits, the D algorithm can cope well with the fault-detecting problem that has been set, which is evidenced by the Figure 5.10 circuit. The 1 $\overline{\text{D}}$ 1 test vector reveals the B s@1 faults for the case of absent fanout in the specific circuit node. If the fanout is found, as in Figure 5.11 for instance, then two fault messages, D and $\overline{\text{D}}$, *quench* each other, and the constant logical unity is found at the circuit output. The bottleneck is called the reconvergent fanout and is associated with the D algorithm's attempts to operate via only one path from the circuit's primary inputs.

The *path-oriented decision-making* (PODEM) algorithm solves the reconvergent fanout problem and permits operation via several rather than one path from the circuit primary input (PI) to its primary output (PO) (multipath sensitization). The algorithm is similar to the base D algorithm, but if an incorrect decision that defies any positive result is made, the algorithm tries to correct it.

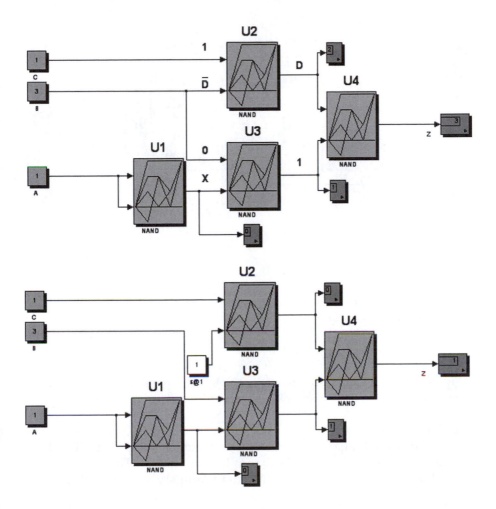

FIGURE 5.10
The circuit where the D algorithm can be used to advantage.

The algorithm contains the following stages.

1. **Objective:** Selection of a circuit node (objective) for its setting to a definite state. The operation starts from the fault origin, and all the remaining circuit nodes are set to state X.

2. **Backtrace:** Return to the corresponding circuit PI and its setting to the state that will turn into the state chosen at stage 1 for the fault origin.

3. **Implication:** The circuit simulation for the calculation of its response to the input influence set at PI at stage 2. If the path sensitization to a certain PO is impracticable in this process, then an attempt is made to invert the PI condition (set at stage 2) and repeat the simulation.

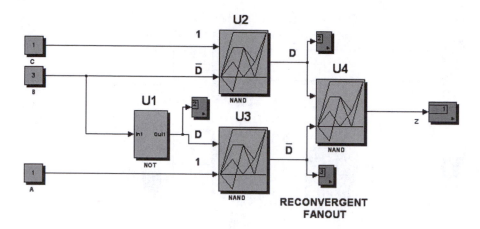

FIGURE 5.11
The case of reconvergent fanout.

4. **D-frontier:** The destination point of the error report "arrival" (D-frontier) is refined, and a backtrace to stage 1 is made if the D-frontier does not reach the targeted PO; otherwise, the algorithm ends, since the target is achieved.

Figure 5.12 presents the model where the following iterations of the above-mentioned four stages of the PODEM algorithm are used.

1. We start from the s@0-type fault sensitization (objective) at the first input of the U2 gate. Then we return to the J and K primary inputs and set J to state 1, while K remains in state X. This stops the D-frontier's advance toward the circuit primary output (PO) Z.

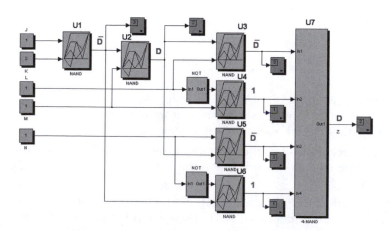

FIGURE 5.12
Illustration of the PODEM algorithm operation.

2. We set K to state 1, which leads to setting the U1 gate output to state 0 and to setting the U6 gate output to state 1. It is evident that the D-frontier's advance to the PO of Z is again absent.

3. To propagate the error report via the U2 gate, we set M to state 1, which results in a D-frontier appearance at the U2 gate output and consequently at the U3 and U5 gate outputs.

4. To propagate the error report via the U5 gate, we set N to state 1, which results in a D-frontier appearance at the U5 gate output and at the third input of the U7 gate.

5. To propagate the error report via the U3 gate, we set L to state 0, which results in the state 1 setting at the U3 gate output and, consequently, to setting of the U4 gate output to state 0. In this process, the path to the Z PO is not sensitized, and, consequently, no D-frontier advance is observed.

As a result, the incorrect decision should be corrected, and L should be set to state 1. Then $\bar{D}$ is set at the U3 gate output, and D at the Z PO. This means the D-frontier has achieved the sought-after goal.

The PODEM algorithm is a further development of the D algorithm. The PODEM algorithm is then developed into a number of highly efficient algorithms described in [1]. Such ATPG algorithms are basic for many commercial ATPG systems.

References

1. Bushnell M.L., Agrawal V.D. Essentials of Electronic Testing for Digital, Memory & Mixed-Signal VLSI Circuits. Kluwer Academic Publishers, Dordrecht, The Netherlands, 2004.

2. Smith M.J.S. Application-Specific Integrated Circuits. Addison-Wesley, Reading, MA, 1997.

3. Pool N.R. Lectures for Course 312EE Advanced Digital Systems (Digital Test). Coventry University, U.K., 1999.

6

Timing Verification

6.1 Logical Determinant Theory [1–4]

6.1.1 Infinite-Valued Logic

Let us consider the infinite point set C = [A,B] as an intercept of A,B straight row. Let us introduce the following logical operation on the C set:

1. Conjunction

$$Y \equiv X_1 \wedge X_2 = \min (X_1, X_2) \qquad (6.1)$$

2. Disjunction

$$Y \equiv X_1 \vee X_2 = \max (X_1, X_2) \qquad (6.2)$$

3. Negation

$$Y \equiv \bar{X} = A + B - X \qquad (6.3)$$

The Equation 6.1 to Equation 6.3 operations generalize the Boolean logic operations and transform into them when C = [0,1] set.

Any arbitrary function, which like its arguments takes the values from the C set and is represented by the superposition of the Equation 6.1 to Equation 6.3 logical operations is called the infinite-valued logic (IVL) function. Such binary functions such as conjunction (Equation 6.1) and disjunction (Equation 6.2) and a monadic function such as negation (Equation 6.3) are the elementary functions of IVL. More complex functions are: n-place conjunction,

$$Y \equiv X_1 \wedge X_2 \wedge \cdots \wedge X_n = \min (X_1, X_2,..., X_n), \qquad (6.4)$$

and n-place disjunction,

$$Y \equiv X_1 \vee X_2 \vee \cdots \vee X_n = \max (X_1, X_2,..., X_n). \qquad (6.5)$$

More complex functions are: elementary n-place conjunction,

$$Y \equiv X_1 \wedge \overline{X}_1 \wedge X_2 \wedge \overline{X}_2 \wedge \cdots \wedge X_n \wedge \overline{X}_n, \qquad (6.6)$$

and elementary n-place disjunction,

$$Y \equiv X_1 \vee \overline{X}_1 \vee X_2 \vee \overline{X}_2 \vee \cdots \vee X_n \vee \overline{X}_n. \qquad (6.7)$$

In the Equation 6.6 and 6.7 expressions some letters can be omitted, but unlike Boolean logic, the IVL elementary conjunctions and disjunctions can contain, along with X_i, its negation, $\overline{X}_i$.

The most complicated IVL functions are the disjunctive normal form (DNF; disjunction of various elementary conjunctions) and the conjunctive normal form (CNF; conjunction of various elementary disjunctions). These functions comprise the Equation 6.1 to 6.7 functions as specific cases.

The implementation of IVL operations using Simulink® blocks is shown in Figure 6.1 (for n = 3).

Any IVL function on any set of arguments assumes the value of an argument or its negation. The IVL function's primary assignment can include the enumeration of all versions of elements ordering with a particular indication of that argument or negation, whose value the function assumes for each version. From this primary assignment of the IVL function, using the superposition of the Equation 6.1 to Equation 6.3 logical operations, we can proceed to its analytic representation. The IVL arbitrary function can be

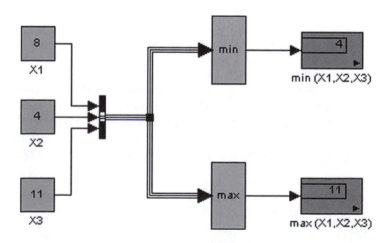

FIGURE 6.1
Implementation of the IVL operations using Simulink blocks (n = 3).

represented in any of its two standard shapes — DNF and CNF — depending on the function analytic assignment.

6.1.2 The IVL Equations and Inequalities

The IVL equation is the equation of the form:

$$F_1(c_1, c_2,..., c_m, X_1, X_2,..., X_n) = F_2(c_1, c_2,..., c_m, X_1, X_2,..., X_n), \qquad (6.8)$$

where F_1 and F_2 are the IVL given functions, $c_1, c_2,..., c_m$ are the known arguments, and $X_1, X_2,..., X_n$ are the unknown arguments. Any set of $X_1, X_2,..., X_n$, where the Equation 6.8 equality turns into identity is called the particular solution of Equation 6.8, whereas the integrity of all particular solutions is called the common solution of Equation 6.8.

The IVL inequality is introduced in similar manner:

$$F_1(c_1, c_2,..., c_m, X_1, X_2,..., X_n) < F_2(c_1, c_2,..., c_m, X_1, X_2,..., X_n). \qquad (6.9)$$

The IVL systems of equations and inequalities can also be discussed. Their major solution technique is the successive partition of their right-hand and left-hand parts, based on the implementation of IVL operations. Such partitioning allows substitution of the initial equation (inequality) for the equivalent combination of the systems of simpler equations and inequalities. The intersection of solutions of the system-involved equations and inequalities is the solution of the equations and inequalities system.

For instance, both parts of the Equation 6.8 solution are represented in the DNF. Then, for both equation parts, the last operation is only conjunction or disjunction. Let us assume that disjunction is the left-hand last operation, that is, that Equation 6.8 is written as:

$$F_1(\mathbf{c}, \mathbf{X}) \vee F_2(\mathbf{c}, \mathbf{X}) = F_3(\mathbf{c}, \mathbf{X}), \qquad (6.10)$$

where $\mathbf{c} = (c_1, c_2,..., c_m)$ is the parameter vector, and $\mathbf{X} = (X_1, X_2,..., X_n)$ is the unknown vector.

By the disjunction definition, Equation 6.10 is equivalent to the combination of two systems of equations and inequalities:

$$\left(\begin{cases} F_1(\mathbf{c}, \mathbf{X}) \geq F_2(\mathbf{c}, \mathbf{X}) \\ F_1(\mathbf{c}, \mathbf{X}) = F_3(\mathbf{c}, \mathbf{X}) \end{cases} \right) \cup \left(\begin{cases} F_1(\mathbf{c}, \mathbf{X}) < F_2(\mathbf{c}, \mathbf{X}) \\ F_2(\mathbf{c}, \mathbf{X}) = F_3(\mathbf{c}, \mathbf{X}) \end{cases} \right) \qquad (6.11)$$

Partition of the left-hand side of the initial Equation 6.10 resulted in each Equation 6.11 equation or inequality being simpler than Equation 6.10, because it has fewer elementary operations. Such a simplification process can proceed further by partitioning the right-hand side of Equation 6.10.

TABLE 6.1

Simple Examples for IVL Equation Solutions

Equation	Systems Combination Using Partitioning of the Equation's Left-Hand Side	System Solutions	Equation Solution
CX = D	$(X \geq C = D) \cup (X = D < C)$	$X \geq D$ with $C = D$ $X = D$ with $C > D$	$X = D$ with $C \geq D$
CX > D	$(X \geq C > D) \cup (D < X < C)$	$X \geq C$ with $> D$ $C > X > D$	$X > D$ with $C > D$
CX < D	$(X \geq C < D) \cup (X < C,D)$	$D > X \geq C$ $X < C,D$	$X \in [A, B]$ with $C < D$ $X < D$ with $C \geq D$
$C \vee X = D$	$(D = X > C) \cup (D = C \geq X)$	$X = D$ with $D > C$ $X \leq D$ with $D = C$	$X = D$ with $D \geq C$

Some simple examples are cited in Table 6.1. The example of the Simulink model for the solution of the IVL equation is shown in Figure 6.2.

6.1.3 Order Logic and Logical Determinants

Let us discuss the $X = \{X_1, X_2,..., X_n\}$ set made of n X_i, $X_i \in [A, B]$ elements. We arrange them in their nondecreasing order:

$$X^1 \leq X^2 \leq \cdots \leq X^n, X^r \in X. \tag{6.12}$$

Now we introduce the single-out operation of an arbitrary order X^r element in X set (the r-operation):

$$Y \equiv F^r (X_1, X_2,..., X_n) = X^r; r = 1, 2,..., n. \tag{6.13}$$

Here, r is called the operation *rank*. The r-operation generalizes the IVL conjunction and disjunction and converts into them with $r = 1$ and $r = n$, respectively.

The r operation over the set of elements results in an element of the same set. An arbitrary function, whose $X_1, X_2,..., X_n$ arguments are the elements of the X set, as the superposition of r operations over X with various r rank values, is called the *order logic function*. The r operation per se is the simplest example of such operation (Equation 6.13).

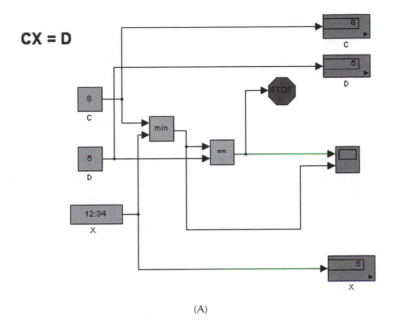

CX = D

(A)

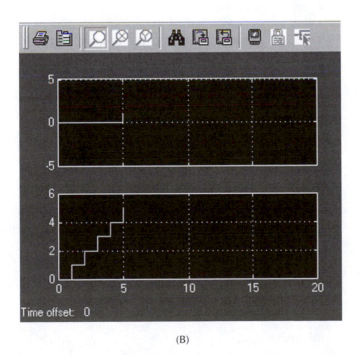

(B)

FIGURE 6.2
The Simulink model for the solution of IVL equation CX = D.

TABLE 6.2

Order Logic Function Y Definition

Ordering of Arguments	Function Value
$X_1 \leq X_2 \leq X_3$	X_3
$X_1 \leq X_3 \leq X_2$	X_2
$X_2 \leq X_1 \leq X_3$	X_3
$X_2 \leq X_3 \leq X_1$	X_1
$X_3 \leq X_1 \leq X_2$	X_2
$X_3 \leq X_2 \leq X_1$	X_1

Any order logic function as the superposition of r operations assumes the value of an argument at any set of $X_1, X_2,..., X_n$ arguments, because the r operation always results in an operation-involved variable.

Here, the order logic function can be set by the enumeration of all n! ordering versions for the $X_1, X_2,..., X_n$ arguments, indicating for each version the X_i argument whose value the function takes. From such primary setting of the order logic function, we can proceed to its analytical representation as the superposition of the IVL operations. For instance, the order logic function $Y = F^3(X_1, X_2, X_3)$ is given in Table 6.2. Its representation can be found by using the IVL operations. According to Table 6.2, the sought-for function can be represented as:

$$Y = \begin{cases} X_1 \text{ with } X_2 \leq X_3 \leq X_1 & \text{or} \quad X_3 \leq X_2 \leq X_1; \\ X_2 \text{ with } X_1 \leq X_3 \leq X_2 & \text{or} \quad X_3 \leq X_1 \leq X_2; \\ X_3 \text{ with } X_1 \leq X_2 \leq X_3 & \text{or} \quad X_2 \leq X_1 \leq X_3. \end{cases}$$

Now, joining all three rows into one by disjunction, we have the required representation:

$$Y = X_1 \vee X_2 \vee X_3.$$

To simplify the expressions of the order logic functions, one can use the equivalent transformations, where the IVL laws are employed. Some specific laws are exclusively pertinent to the order logic:

1. The tautology law:

$$F^r (X, X,..., X) = X \tag{6.14}$$

2. The commutative law:

$$F^r (X_1, X_2,..., X_n) = F^r (X_{i1}, X_{i2},..., X_{in}) \tag{6.15}$$

$(X_{i1}, X_{i2},..., X_{in}$ – any commutation of the $X_1, X_2,..., X_n$ arguments)

3. The distributive law and its special cases:

$$F^r \left[\Phi^{q1}(X_1, X_2,\ldots, X_n), \Phi^{q2} (X_1, X_2,\ldots, X_n),\ldots, \Phi^{qn} (X_1, X_2,\ldots, X_n) \right]$$

$$= \Phi^{qr} (X_1, X_2,\ldots, X_n), (q_1 < q_2 < \cdots q_p; 1 \leq r \leq p); \tag{6.16}$$

and its special cases (here, the minimum and maximum finding operations are performed for all $1 \leq i \leq n$ index values):

$$\min F^{ri} (X_1, X_2,\ldots, X_n) = F^{\min ri} (X_1, X_2,\ldots, X_n)$$

$$\max F^{ri} (X_1, X_2,\ldots, X_n) = F^{\max ri} (X_1, X_2,\ldots, X_n) \tag{6.17}$$

Now we will discuss the X_q set that is composed of q nonintersecting subsets $(X_{i1}, X_{i2},\ldots, X_{im_i})$ with $X_{ij} \in [A, B]$ elements ordered according to the condition

$$X_{i1} \leq X_{i2} \leq \cdots \leq X_{im_i}; i = 1, 2,\ldots q. \tag{6.18}$$

The number of such set elements $n = \Sigma m_i$ (the summation is made from $1 \leq i \leq q$).

It is evident that the X_q set is a partially ordered set that can be written as the q-th order *quasi-matrix* (the quasi matrix rows are the ordered subsets):

$$X_q = \begin{Vmatrix} X_{11} \cdots X_{1m_1} \\ \cdots\cdots\cdots\cdots \\ X_{q1} \cdots X_{qm_q} \end{Vmatrix} = \|X_{ij}\|; \tag{6.19}$$

$$i = 1,\ldots q; \quad j = 1,\ldots, m_i.$$

Quasi-matrix differs from common matrix by a dissimilar number of row elements and by the element ordering in each row, according to Equation 6.18. The above-specified disordered $X = \{X_1, X_2,\ldots, X_n\}$ set is a special case of Equation 6.19, each with single-element rows. Thus, the disordered set of X is written as the *column-matrix*:

$$X_q = \begin{Vmatrix} X_1 \\ \cdots \\ X_n \end{Vmatrix}. \tag{6.20}$$

The quasi-matrix alternative special case is the *row matrix:*

$$X = \|X_1, X_2,\ldots, X_n\|. \tag{6.21}$$

It is clear that in this case the X set is completely ordered.

For the partially ordered X_q set, prescribed by its quasi-matrix of Equation 6.19, the r operation is inserted, similarly to the above-described step for the completely ordered set of Equation 6.12:

$$Y \equiv F^r (X_{11}, X_{12}, \dots, X_{qm_q}) = X^r \; ; \; r = 1, 2, \dots, n. \tag{6.22}$$

This function separates the required order element X^r out of X_q and is called the *r rank q-th order logical determinant (LD)* of X_q quasi-matrix:

$$X^r_q = \begin{vmatrix} X_{11} \cdots X_{1m_1} \\ \cdots\cdots\cdots\cdots \\ X_{q1} \cdots X_{qm_q} \end{vmatrix}^r = \left| X_{ij} \right|^r ; \tag{6.23}$$

$$i = 1, \dots q; \quad j = 1, \dots, m_i; \quad r = 1, 2, \dots, n.$$

The LD special cases are column determinant and row determinant. Column determinant,

$$X^r_n = \begin{vmatrix} X_1 \\ \cdots \\ X_n \end{vmatrix}^r \tag{6.24}$$

corresponds to the column matrix of Equation 6.20, whereas row determinant,

$$X^r_1 = |X_1, X_2, \dots, X_n|^r = X_r, \tag{6.25}$$

corresponds to the quasi-matrix of Equation 6.21.

The X^r_q logical determinant of the X_q quasi-matrix is the matrix numerical characteristic, much like the common determinant is the characteristic of a common quadratic matrix.

The LD is the generalization of the common r function of Equation 6.13 for the case of a partially ordered set of arguments. Thus, the LD and the LD superposition can be assigned by specifying such X_{ij} element for each ordering version of the $X_{11}, X_{12}, \dots, X_{qm_q}$ elements that its value is taken by the function being studied. Then, we can proceed to their analytical representation using the IVL operations.

The LD exposure means the LD analytic representation as the IVL function, via the values of its arguments. The r-th rank LD with n elements is expressed as the DNF,

$$X^r_n = \begin{vmatrix} X_1 \\ \cdots \\ X_n \end{vmatrix}^r = \max(X_{i_1} \cdots X_{i_n} - r + 1); \quad X_{i_k} \in \{X_1, X_2, \dots, X_n\}; \tag{6.26}$$
$$(i_1 \neq \cdots \neq i_{n-r+1})$$

or as the following CNF:

$$X^r_n = \begin{vmatrix} X_1 \\ \cdots \\ X_n \end{vmatrix}^r = \min(X_{i_1} \vee \cdots \vee X_{i_n} - r + 1); \quad X_{i_k} \in \{X_1, X_2, \ldots, X_n\}; \quad (6.27)$$
$$(i_1 \neq \cdots \neq i_r)$$

The common infinite r-th rank q-th order LD is expressed as the DNF,

$$X^r_q = \begin{vmatrix} X_1 \cdots X_{1i_1} \cdots \\ \cdots\cdots\cdots\cdots\cdots \\ X_{q1} \cdots X_{qi_q} \cdots \end{vmatrix}^r = \max(X_{1i_1} \cdots X_{qi_q}); \quad (6.28)$$
$$(\Sigma i_s = r + q - 1; \ s = 1, 2, \ldots, q)$$

whereas the common finite r-th rank n-th order LD is expressed as

$$X^r_q = \begin{vmatrix} X_{11} \cdots X_{1m_1} \\ \cdots\cdots\cdots\cdots \\ X_{q1} \cdots X_{qm_q} \end{vmatrix}^r = \max\left(X^{m_1}_{1i_1} \cdots X^{mq}_{qi_q}\right). \quad (6.29)$$
$$(\Sigma i_s = r + q - 1; s = 1, 2, \ldots, q)$$

Here and below, the $X^{m_k}_{ki_k}$ entry means that the X_{ki_k} element is excluded from all those conjunctions (disjunctions) where the $i_k > m_k$ condition is formally obtained from the condition on Σi_s.

Example 1: Exposure of the LD Column from Equation 6.26:

$$X^r_3 = \begin{vmatrix} x_1 \\ x_2 \\ x_3 \end{vmatrix}^r = \begin{cases} x_1\, x_2\, x_3, & r = 1; \\ x_1\, x_2 \vee x_1\, x_3 \vee x_2\, x_3, & r = 2; \\ x_1 \vee x_2 \vee x_3, & r = 3. \end{cases}$$

Example 2: Exposure of the Common Second-Order LD from Equation 6.29

$$X^r_2 = \begin{vmatrix} x_{11} & x_{12} \\ x_{21} & x_{22} \end{vmatrix}^r = \begin{cases} x_{11}x_{12}, & r = 1; \\ x_{11}x_{22} \vee x_{12}x_{21}, & r = 2; \\ x_{12}x_{22} \vee x_{11} \vee x_{21}, & r = 3; \\ x_{12} \vee x_{22}, & r = 4. \end{cases}$$

6.1.4 Exposure of Large LDs

Exposure of large LDs (LDs with a large number of elements) from the explicit formulas is an extremely labor-consuming process. For such cases the decomposition of the source LD into smaller LDs (units) in conformity with the decomposition rules is quite expedient [1, 2].

Decompositions are the basis for the hierarchical procedures of LD exposure. In such procedures, the decomposition of the calculated LD into the units, that is, the lower-order LDs, is first executed. Then, the LDs obtained are further decomposed into the LDs of much lower order, and so on until the source LD is expressed by the first-order LD, that is, the x_{ij} elements. The procedure's laboriousness and the complexity of the LD obtained are determined by the LD decomposition scenario. The most efficient method is to decompose the LD into units so that at each step the LD is subdivided into two similar-sized LDs. The thus obtained LD expressions have the complexity of:

$$N^r_q = r^2 \times (q - 1) + 2r - 1 \tag{6.30}$$

and

$$N^r_q = k \times r \times n, \quad k \le 2 \quad (k - constant). \tag{6.31}$$

The Equation 6.31 estimate is based on the assumption that all q rows of the LD contain the same number of m elements (the overall number of LD elements is n = m q). Use of dichotomic unit decompositions exposes large LDs with acceptable calculation complexities. If very large LDs must be exposed, it is expedient to use their approximate exposure based on the bilateral analytical estimates of LD value [1, 2].

6.1.5 Probabilistic Calculations in IVL

If IVL functions that have random variables as arguments should be considered, such IVL distributions and moments of functions should be found whose arguments are distributed according to the known statistical laws. Some of the results obtained in the probabilistic IVL can be found in [1, 2].

6.2 Digital Circuit Dynamics [1–4]

6.2.1 Statistical and Dynamical Analysis of Combinational Circuits

Let the synchronous combinational circuit with n inputs and m outputs be given. We must find the function it implements, that is, define the outlook of the circuit output set transformed out of the circuit input set.

The most natural way to solve this problem is the following permutation technique. We hold a certain set fixed at the circuit inputs. Now we highlight all gates, whose inputs are only circuit inputs, and classify them as the circuit's first cascade. Having calculated the Boolean function values for these gates, we find the effective set of signals at the outputs of the circuit's first cascade. Then we highlight all gates, whose inputs are the circuit inputs and the first cascade gate inputs, and classify them as the circuit's second cascade. Thus, the set of signals for the second cascade inputs is also identified. Now, having calculated the Boolean function values for the second-cascade gates, we find the set of signals at the outputs of the circuit's second cascade. The subsequent procedures are executed similarly for all circuit cascades. As no closed loops are present in a combinational circuit, the above-described analytic procedure demands only a finite number of steps and results in the set of signals at the circuit outputs, that is, the circuit output set as a response to the circuit's corresponding fixed input set. Analytical procedures for alternative input sets are similar to those just described and are executed following the circuit's cascading.

All sets are given in a unified graphic shape. Step k of the procedure is the search for logical expressions related to the variable outputs of the circuit k cascade elements as the functions of the circuit inputs, by substituting the Boolean function arguments of the k cascade elements for logical expressions of the corresponding (k − 1) cascade outputs found at the previous step. The logical expressions are further simplified by their equivalent, Boolean-algebra-based conversion (tautological, commutative, combinatorial, negation [de Morgan], absorption, double negation, sewing, and deorthogonalization).

Following the above procedure, we obtain the expressions of the circuit-implemented Boolean functions in terms of input variables. Statistical analysis of synchronous combinational circuits assumes that each gate at any moment of time implements a Boolean function of the values of its input variables at the same moment:

$$y(t) = f(x_1(t), x_2(t),\ldots, x_n(t)). \tag{6.32}$$

When the gates are inertia-free and when all circuit input variables change simultaneously, each output changes its value at the same moment of time — from the initial value (corresponding to the circuit's initial input set) to the final value (corresponding to the circuit's final input set). In these conditions, statistical analysis gives the exhaustive description of combinational circuit operation.

In practice, the changes in the circuit input set frequently occur with time-mismatched changes of various input variables, whereas the gates have inertia. In this process the combinational circuit is exposed to a number of successively altering input sets so that each of them, by passing through the circuit, is further subdivided into several more sets (owing to the gate's inertia-mismatched values). As a result, signal variations at each circuit output and its inner nodes do not occur immediately but rather generate a certain complex commutational process. Detection of such a dynamic processes is the concern of the combinational circuit's dynamical analysis.

We will examine the inertial combinational circuits that differ from the inertia-free ones. The changes in the inertial combinational circuit input set involves the nonsimultaneous changes in various set-comprising input variables. Hence, we speak here of the set of input 0 and 1 sequences, acting independently at various combinational circuit inputs rather than on the sequence of the circuit input sets. In addition, some auxiliary elements such as delays and filters can be found in combinational circuits. The delay element changes the input signal time shift by the constant value of τ. In doing this, the output signal shape is similar to that of the input signal. The filter executes the same operation as the delay element, but it hinders the output of input signal changes that have time intervals of at least τ. Owing to such filtering of the frequent signal changes, the filter output process can differ from the input process.

Each inertial gate can be represented as a successive connection of the inertia-free gate, implementing the Equation 6.32 inertia-free transformation and the τ delay element. Generally, shaping the delay value depends on the type of gate output signal variation: $(0 \rightarrow 1$ or $1 \rightarrow 0)$. The values that characterize the delay element and the filter and the moments of combinational circuit input variable changes are called the time parameters [1, 2].

6.2.2 Switching Dynamic Processes in the System [1,2]

The study object of the combinational circuit dynamic analysis is not a Boolean variable with two values of (0, 1) but its changes, in conjunction with variations in the changes. For further presentation, the following definitions should be introduced: 1 is a constant signal equal to a logical unity at a certain time interval. 0 is a constant signal equal to a logical zero at a certain time interval. $1'_a$ is signal variation at the moment:

$$1'_a = \begin{cases} 0, & t < a; \\ 1, & t \geq a; \end{cases} \tag{6.33}$$

$0'_b$ is signal variation at the b moment:

$$0'_b = \begin{cases} 1, & t < b; \\ 0, & t \geq b; \end{cases} \tag{6.34}$$

1(a, b) is the $1'_a$ $0'_b$ pulse:

$$1(a,b) = \begin{cases} 1, & a \leq t < b; \\ 0, & t < a \quad \text{or} \quad t \geq b; \end{cases}$$ (6.35)

0(a, b) is the $0'_a$ $1'_b$ pause:

$$0(a,b) = \begin{cases} 0, & a \leq t < b; \\ 1, & t < a \quad \text{or} \quad t \geq b. \end{cases}$$ (6.36)

Equations 6.35 and 6.36 with a = b become

$$1(a, a) \equiv 0; \quad 0(a, a) \equiv 1.$$ (6.37)

The signal variations in Equations 6.32 and 6.33 can also be considered as a pulse (a pause) at $(-\infty, a)$, (a, ∞) intervals:

$$1'_a = 1(a, \infty) = 0(-\infty, a);$$
$$0'_b = 0(a, \infty) = 1(-\infty, a).$$ (6.38)

The x(t) switching dynamic process is a sequence of pulses and pauses at the $(-\infty < t < \infty)$ time interval, which contains a finite number of pulses and pauses at any finite subinterval of a given interval.

Let x(t) be an arbitrary switching process differing from those identical to 0 or to 1. a_x is the point of the first signal variation in x(t), and b_x is the point of the last signal variation in x(t). The x(t) process with a_x finite value (b_x finite value) is called the left (right) finite process, whereas the x(t) process with an $a_x = -\infty$ ($b_x = \infty$) value is the left (right) infinite process. The process that is both left and right finite (infinite) is called the finite (infinite) process.

The x_0 value of the x(t) left finite process with $t < a_x$ is called the initial x(t) value. The x(t) starts from pulse (pause) if $x_0 = 0$ ($x_0 = 1$). The x_∞ value of the x(t) right finite process with $t > a_y$ is called the x(t) terminal value, taking x(t) terminating in pulse (pause) if $x_\infty = 0$ ($x_\infty = 1$). The x(t) right finite and the y(t) left finite processes are called the time nonintersecting processes if $b_x \leq a_y$.

The switching process length is the overall amount of signal variation therein. For a vector process, the corresponding value is the length vector. The process lengths (vector process components) are the components of such a vector. The switching process is called complex if its length is more than or equal to two. In the opposite case, it is called a simple switching (or a degenerate switching process).

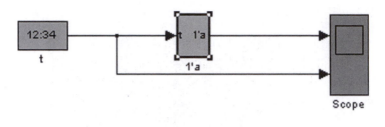

FIGURE 6.3A
A model of the 0 → 1 transition.

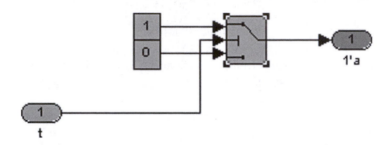

FIGURE 6.3B
Internal structure of the $1'_a$ unit for the model of the 0 → 1 transition.

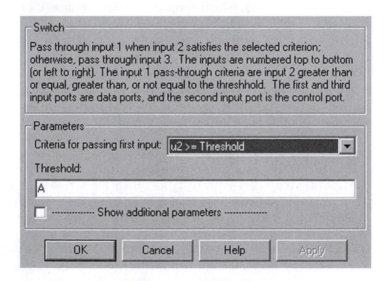

FIGURE 6.3C
Assignment of the *a* parameter for the $1'_a$ unit in the model of the 0 → 1 transition (its specific value in our case is a = 8) is given before the simulation in the main window of the MATLAB system.

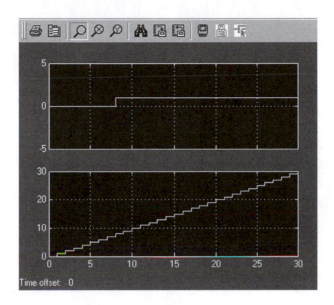

FIGURE 6.3D
Input and output of the model of the 0 → 1 transition.

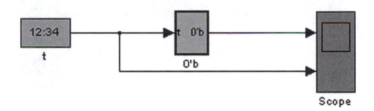

B=15

FIGURE 6.4A
The model of the 1 → 0 transition.

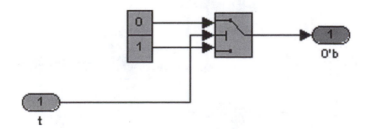

FIGURE 6.4B
Internal structure of the $0'_b$ unit for the model of the 1 → 0 transition. Assignment of the *b* parameter for the $0'_b$ unit in the model of the 0 → 1 transition (its specific value for our case is b = 8) is given in the manner similar to the *a* parameter assignment in the previous model (see Figure 6.3).

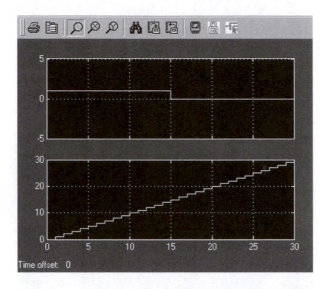

FIGURE 6.4C
Input and output of the model of the 1 → 0 transition.

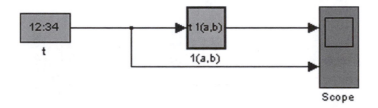

A=8 B=15

FIGURE 6.5A
The pulse model.

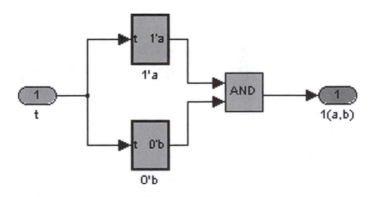

FIGURE 6.5B
Internal structure of the 1(a, b) unit for the pulse model.

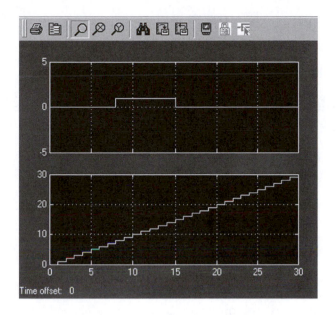

FIGURE 6.5C
The pulse model input and output.

We can show a number of Simulink models for the transitions between the logic states of signals, pulses, pauses, and switching processes:

1. The $0 \rightarrow 1$ transition model (Figure 6.3A–D)
2. The $1 \rightarrow 0$ transition model (Figure 6.4A–C)
3. The pulse model (Figure 6.5A–C)
4. The pause model (Figure 6.6A–C)
5. The pulse-pause-containing switching process model (Figure 6.7A–C)
6. The two-pulse-pause-containing switching process model (Figure 6.8A–C)

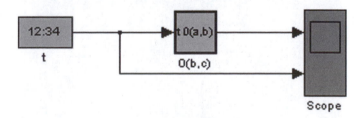

FIGURE 6.6A
The pause model.

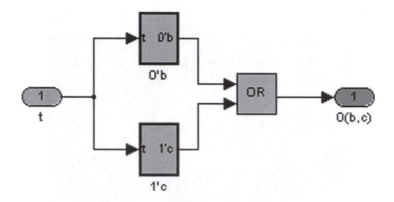

FIGURE 6.6B
Internal structure of the 0(b, c) unit for the pause model.

The actions at the gate inputs are given as certain $x_i(t)$ switching processes. The gate output $y(t)$ response to said actions is called the gate dynamic (transition) process, corresponding to these actions. When the gate is investigated in isolation (i.e., outside the circuit), the process at the output of the gate's inertia-free section is specified as its dynamic process. The properly dynamic process in the gate is obtained from $y(t)$ by the τ time shift, without any shape changes.

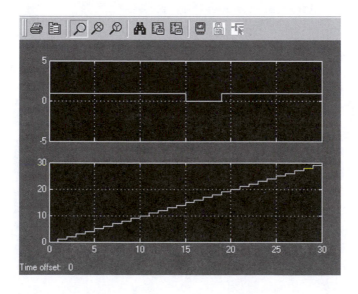

FIGURE 6.6C
The pause model input and output.

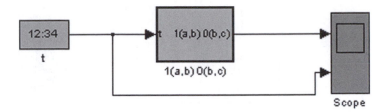

A=8 B=15 C=19

FIGURE 6.7A
The model of the pulse-pause-containing switching process.

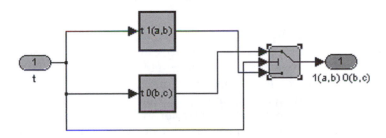

FIGURE 6.7B
Internal structure of the 1(a, b)0(b, c) unit for the model of the pulse-pause-containing switching process.

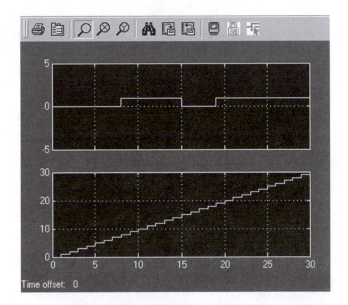

FIGURE 6.7C
Input and output for the model of the pulse-pause-containing switching process.

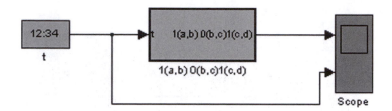

A=8 B=15 C=19 D=23

FIGURE 6.8A
The model of the two-pulse-pause-containing switching process.

The actions at memory-free combinational circuit inputs are given as certain $x_i(t)$ switching processes. The integrity of the output responses from all combinational circuit gates is called the combinational circuit dynamic process, corresponding to all these input actions.

Set the task of finding the dynamic process's symbolic shape at any combinational circuit output, based on the input switching processes of the same kind [1,2]. As long as the dynamic process at all combinational circuit outputs is sought by the same algorithm, we can restrict the discussion to the single-output combinational circuits. The problem of the dynamic process finding in any combinational circuit's internal node is also reduced to the same scenario.

The combinational circuit behavior is called proper if it coincides with the behavior that can be found when all gates are inertia-free: if a simple vector switching (and all components change at the same moment of time) is operative at the combinational circuit inputs, then a dynamic process at each combinational circuit input is a simple switching (a degenerate

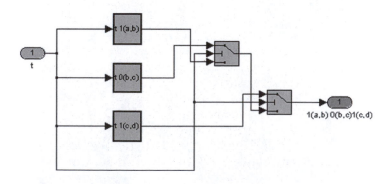

FIGURE 6.8B
Internal structure of the 1(a, b)0(b, c)1(c, d) unit for the model of the two-pulse-pause-containing switching process.

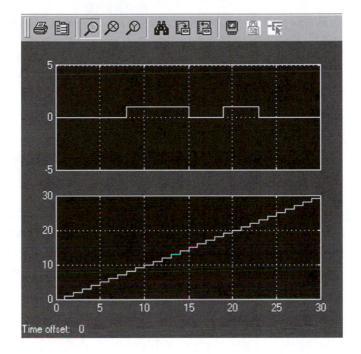

FIGURE 6.8C
Input and output for the model of the two-pulse-pause-containing switching process.

switching process) that takes place at the same moment of time. Such a definition permits finding the exact location of a fault during the combinational circuit operation.

Let a simple vector switching act on the combinational circuit inputs. The single-input (multi-input) combinational circuit operation is faulty if the input (inputs) dynamic transition process, corresponding to this action, is the complex scalar (vector) switching process.

Thus, the faults in combinational circuit operation result from (a) diverse times of changes in the combinational circuit input variables, constituting the given change in the input set, and (b) gate inertia (signal delays).

Two switching processes are equal if their signal variation moments, as well as the changes per se, coincide.

Two expressions that describe the switching process with graphically given moments of signal variations are thought equivalent if both switching processes are equal at any numerical concretization of said moments. The switching process is recorded either as a sequence of signal variations or as a pulse-pause sequence. To save room, the signal initial and final constant values are omitted, and the points of signal intermediate variations are indicated only once, since the point of the pulse (pause) end coincides with the point of pause (pulse) beginning. For instance, the switching process in Figure 6.8 can be described as

$$x(t) = 1'_a \, 0'_b 1'_c 0'_d$$

or

$$x(t) = 1(a, b)0(-,-)1(c, d).$$

The combinational circuit dynamic analysis is also made by the permutation technique, similar to the technique employed in statistical analysis. The difference is as follows:

1. The dynamical analysis problem's dimensionality is much higher because the combinational circuit responses to all potential changes in input sets should be found (in statistical analysis, we need only find the combinational circuit response to all potential input sets).
2. The switching process length, by passing through gate, can increase (the process ramification effect).

Hence, it follows that the combinational circuit dynamical analysis is also reduced to the finding of dynamical processes in a typical gate, where input impacts are the switching processes of sufficient length.

The unit dynamical process in high-dimensional combinational circuits is based on the LD mathematical aids. However, in this process definite aggregates of such parameters described by quasi-matrices and LDs are examined, instead of individual time parameters of a combinational circuit and its input actions. This provides a possibility of combinational circuit unit-wise description and enhances the visibility of analyzed dynamical processes. Such an approach is similar to the matrix analysis of linear systems; the quasi-matrices and LDs play the role of unit description parameters, much as common matrices and determinants in linear dynamical systems.

6.2.3 Dynamical Processes in Multi-Input Gates Following Arbitrary Length Impacts [1–4]

Let us discuss an arbitrary vector switching process,

$$x(t) = \{ x_1(t), x_2(t),\dots, x_n(t)\}, \tag{6.39}$$

where $x_i(t)$ are common switching processes. For certain definiteness, we think they start and finish by

$$x_1(t) = 1(a_{11}, b_{11})0(-,-)1(a_{12}, b_{12})\dots1(a_{1m_1}, b_{1m_1}),$$

$$\dots\dots\dots\dots\dots\dots\dots\dots\dots\dots\dots\dots \tag{6.40}$$

$$x_n(t) = 1(a_{n1}, b_{n1})0(-,-)1(a_{n2}, b_{n2})\dots1(a_{nm_n}, b_{nm_n})$$

pulses.

Let the Equation 6.39 process be the input action of an n-input inertia-free gate that implements the symmetrical Boolean function, that is, the function that undergoes no changes by arbitrary renumeration of arguments.

The following theorem is given in [1, 2]. The totality of the arbitrary switching processes of Equation 6.40, acting at the inputs of any gate of the symmetrical Boolean function, can be replaced (without altering the gate response) by the totality of $M = \Sigma m_i$ $(i = 1,\ldots, n)$ pulses of $1(a^r, b^r)$; $r = 1, 2,\ldots, M$, whose existence intervals are expressed by the Equation 6.40 process parameters, using LD (Equation 6.41):

$$a^r = \begin{vmatrix} a_{11} \cdots a_{1m_1} \\ \cdots\cdots\cdots\cdots \\ a_{n1} \cdots a_{nm_n} \end{vmatrix}; \tag{6.41}$$

and

$$b^r = \begin{vmatrix} b_{11} \cdots b_{1m_1} \\ \cdots\cdots\cdots\cdots \\ b_{n1} \cdots b_{nm_n} \end{vmatrix}; \quad r = 1, 2,\ldots, M;$$

and the pulses are arranged so that the larger number pulse begins (ends) after the smaller number pulse. In this process a pulse action is independent of its feeding input.

The n-input disjunction response to Equation 6.40 impacts

$$y_1(t) = 1(a^1, b^1\, a^2)0(-,-) \cdots 1(a^{M-1}, b^{M-1}\, a^M)0(-,-)1(a^M, b^M), \tag{6.42}$$

where a^r and b^r are determined from Equation 6.41.

The expression shows that the disjunctor response to those input impacts that start with and end in pulses is generally the sequence that starts with and ends in pulses and that contains the number of pulses equal to (or smaller than) the sum of similar numbers for the input sequences.

The Equation 6.42 response to the Equation 6.40 basis actions that start with and end in pulses can be used for finding the n-input response to any alternative impacts. Therewith, the impacts that start with (end in) pulses are examined, but those in which the first pulse starts at $t = -\infty$ moment (the last pulse ends at $t = \infty$ moment).

Thus, it is evident that the disjunctor response to the Equation 6.40 impacts can contain up to M pulses separated by pauses [5–7]:

$$y_1(t) = \Sigma 1(a^j, b^j\, a^{j+1}), \quad (1 \le j \le M) \tag{6.43}$$

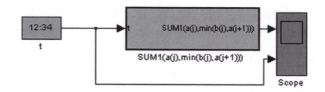

A(1)=8 A(2)=10 A(3)=19 B(1)=15 B(2)=20 B(3)=23

FIGURE 6.9A
The model of a three-input disjunctor (version 1) for M = 3.

(the j pulse is nondegenerate if the $a^j < b^j a^{j+1}$ condition is met; an actually nonexisting M + 1 element is not examined).

The model of a three-input disjunctor (version 1) for M = 3 is shown in Figure 6.9A–G. The model of a three-input disjunctor (version 2) is shown in Figure 6.10A–H.

The n-input conjunctor response to the Equation 6.40 impacts looks like

$$y_2(t) = 1(b^1a^n, b^1)0(-,-)1(b^2a^{n+1}, b^2)\dots1(b^{M+1-n} a^M, b^{M+1-n}), \qquad (6.44)$$

where a^r and b^r are defined from Equation 6.41.

Thus, it is evident that the conjunctor response to the Equation 6.40 impacts can contain up to M + 1 − n pulses separated by pauses [5–7]:

$$y_2(t) = \Sigma1(b^j a^{n+j-1}, b^j), (1 \le j \le M + 1 - n). \qquad (6.45)$$

Hence, the $y_2(t)$ calculation is reduced to the definition of pulse boundaries. The j-th pulse does exist and has the finite length when the $a^{n+j-1} \le b^j$ condition

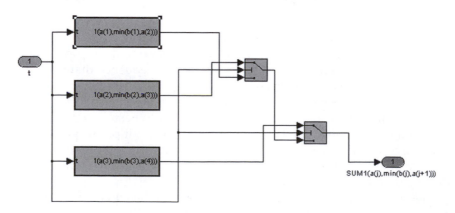

FIGURE 6.9B
The structure of the model of a three-input disjunctor (version 1) for M = 3.

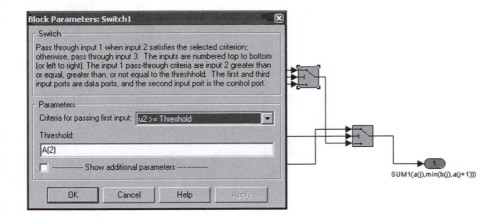

FIGURE 6.9C
Parameter setting for the model switch units (upper level).

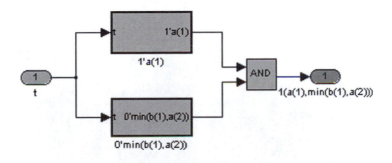

FIGURE 6.9D
Internal structure of 1(a(1), min(b(1), a(2))) unit.

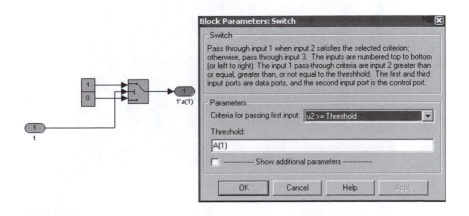

FIGURE 6.9E
Internal structure of 1'a(1) unit.

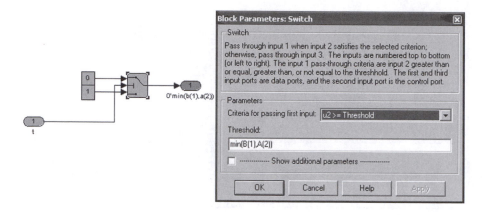

FIGURE 6.9F
Internal structure of 0'min(b(1),a(2)) unit.

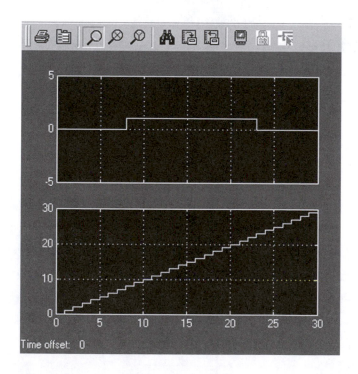

FIGURE 6.9G
The three-input disjunctor response to the impacts described by the Figure 6.9A parameters.

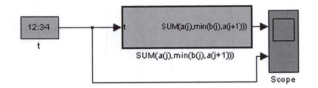

$$A(1)=8 \; A(2)=16 \; A(3)=19 \; B(1)=15 \; B(2)=18 \; B(3)=23$$

FIGURE 6.10A
The model of a three-input disjunctor (version 2) for M = 3.

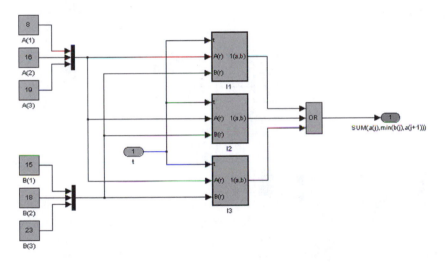

FIGURE 6.10B
The structure of the model of a three-input disjunctor (version 2) for M = 3.

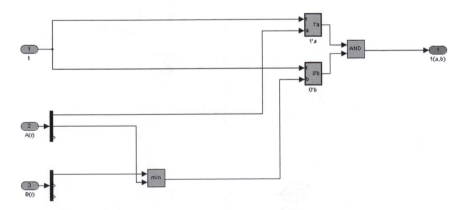

FIGURE 6.10C
The structure of the I1 unit.

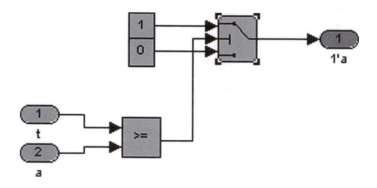

FIGURE 6.10D
The structure of the 1′a unit.

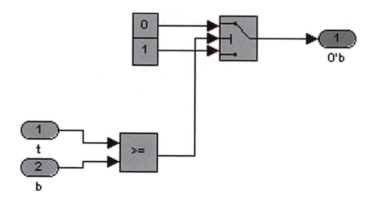

FIGURE 6.10E
The structure of the 0′b unit.

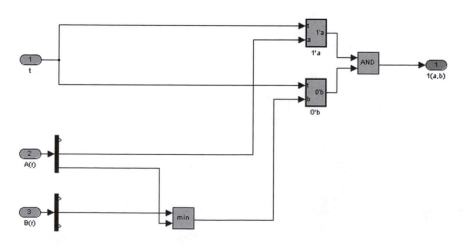

FIGURE 6.10F
The structure of the I2 unit.

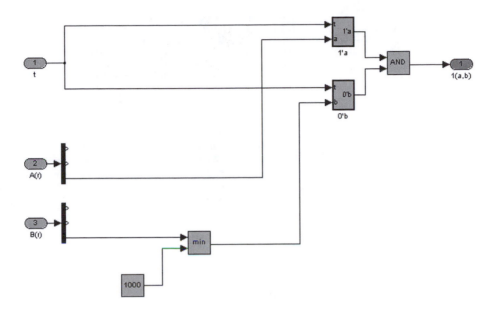

FIGURE 6.10G
The structure of the I3 unit.

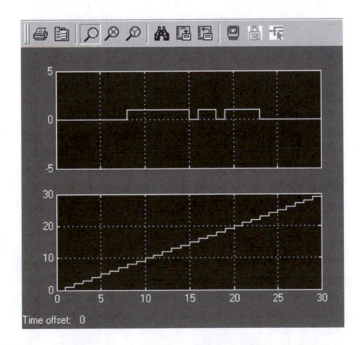

FIGURE 6.10H
The three-input disjunctor response to the impacts described by the Figure 6.10A parameters.

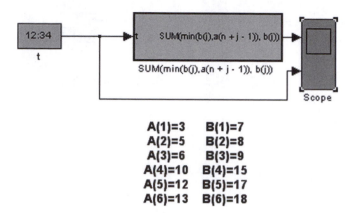

A(1)=3   B(1)=7
A(2)=5   B(2)=8
A(3)=6   B(3)=9
A(4)=10 B(4)=15
A(5)=12 B(5)=17
A(6)=13 B(6)=18

FIGURE 6.11A
The model of the n-input conjunctor for M = 6, n = 3.

is met. The model of the n-input conjunctor for M = 6, n = 3 is given in Figure 6.11A–G.

Let us denote the fundamental symmetrical p index function of n variables as f_n^p. By definition, $f_n^p = 1$ if exactly p variables (irrespectively of their type) are equal to 1. The response of the f_n^p implementing gate to the Equation 6.39 impacts looks like

$$f_n^p (t) = 1(b^1 a^p, b^1\, a^{p+1})0(-,-)1(b^1 \vee b^2 a^{p+1}, b^2 a^{p+2}) \cdots 1(b^{M-p-1} \vee b^{M-p}\, a^{M-1},$$
$$b^{M-p}\, a^M)\, 0(-,-)\cdots 1(b^{M-p} \vee b^{M-p+1}\, a^M, b^{M-p+1}), \tag{6.46}$$

where a^r and b^r are found from Equation 6.41. It is evident that the gate response to the Equation 6.40 impacts can contain up to $M - p + 1$ pulses separated by pauses [5–7]:

$$f_n^p (t) = \Sigma 1(b^{j-1} \vee b^j a^{j+p-1}, b^j a^{j+p}),\ (1 \le j \le M - p + 1). \tag{6.47}$$

The $f_n^p (t)$ calculation is reduced to the definition of boundaries for all potential pulses. The j-th pulse does exist and has a finite length when the

$$b^{j-1} \vee b^j a^{j+p-1} < b^j a^{j+p} \Rightarrow b^{j-1} < b^j a^{j+p}$$

condition is met (the actually nonexisting $b^0 \equiv -\infty$ and $a^{M+1} \equiv \infty$ elements are not examined).

The model of fundamental symmetric function of $f_4^0 (t)$ (M = n = 4) is given in Figure 6.12A–H.

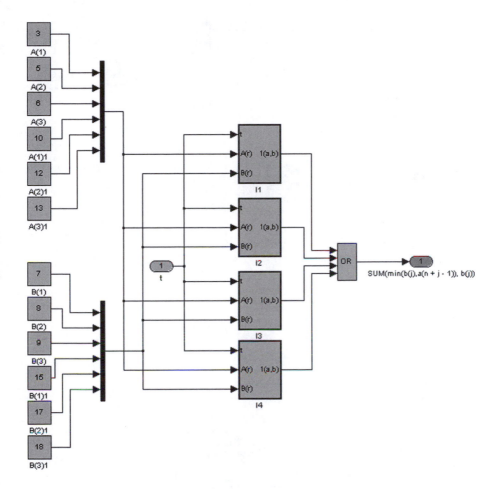

FIGURE 6.11B
The structure of the model of the n-input conjunctor for M = 6, n = 3.

6.2.4 Logical Determinants and Sorting [5–7]

Calculation of gate responses to the given input impacts is subdivided into two stages:

1. Separate ordering of two sets:

$$\{a_1,..., a_n\} \rightarrow \{a^1,..., a^n\}$$

$$\{b_1,..., b_n\} \rightarrow \{b^1,..., b^n\}$$

2. Joint ordering of $\{a^1,..., a^n\}$, $\{b^1,..., b^n\}$ sets according to Equations 6.42 to 6.47 and analogous formulas

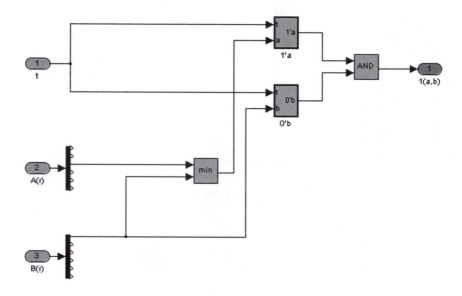

FIGURE 6.11C
The structure of the I1 unit.

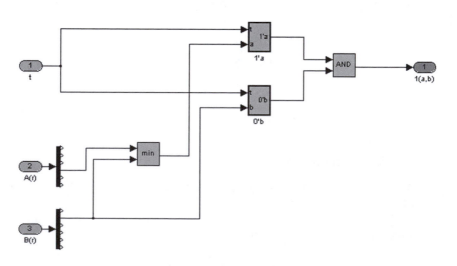

FIGURE 6.11D
The structure of the I2 unit.

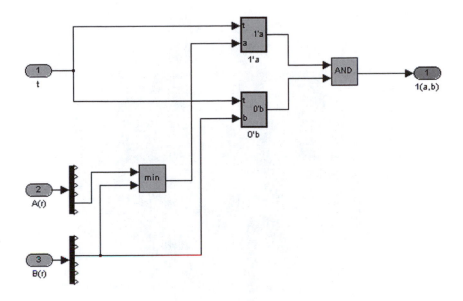

FIGURE 6.11E
The structure of the I3 unit.

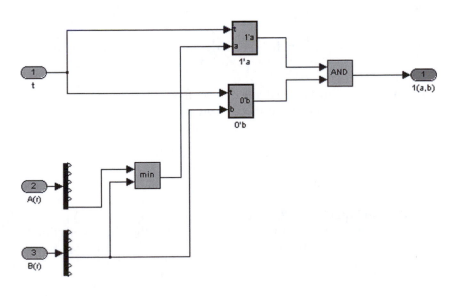

FIGURE 6.11F
The structure of the I4 unit.

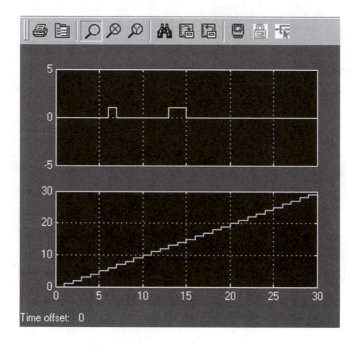

FIGURE 6.11G
The three-input conjunctor response to the impacts described by the Figure 6.11A parameters.

In this context it is evident that the complexity of gate response calculation algorithms is dictated by the complexity of the applied ordering (sorting) algorithms.

The LD facility can be used for the analysis and synthesis of sorting algorithms in data arrays. Each LD is represented by the search function:

$$a^r = \max\left(a_{1i_1}^{m_1}, a_{2i_2}^{m_2}, \ldots a_{ni_n}^{m_n}\right).$$

$$\Sigma i_s = r + n - 1 \ (1 \leq s \leq n). \tag{6.48}$$

According to Equation 6.48, the a^r search algorithm is in the $A_n = Q_1 \cup Q_2 \cup \cdots \cup Q_n$ set, consisting of the finite n number of the nonintersecting $Q_i = \{a_{i1}, \ldots, a_{im_i}\}$, $i = 1, 2, \ldots, n$ subsets, with $a_{ij} \in \{-\infty, \infty\}$ numerical elements arranged according to

$$a_{i1} \leq a_{i2} \leq \cdots \leq a_{im_i} \ ; i = 1, 2, \ldots n;$$

all possible sets containing one element of each n subset should be isolated in such a manner that the sum of second indices for i_s elements of each set equals $r + n - 1$ (if the condition of Σi_s renders $i_k > m_k$ for the a_{ki_k} element of the k-th subset, then the k-th subset element is lacking in the set). A minimal

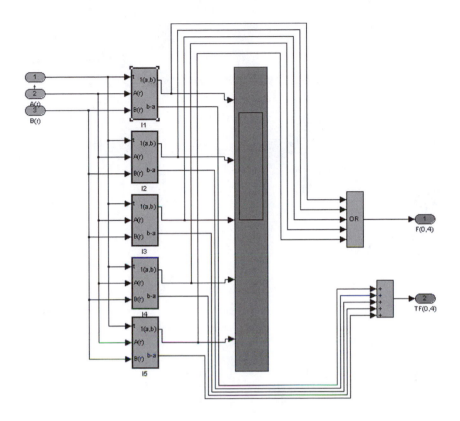

FIGURE 6.12A
The model of the fundamental symmetrical function, f_4^0 (t) (M = n = 4).

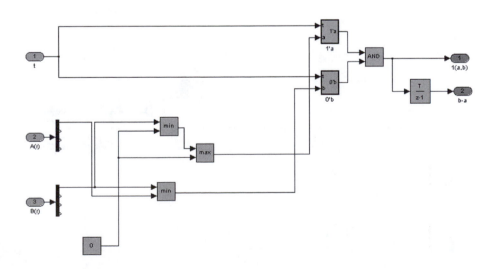

FIGURE 6.12B
The structure of the I1 unit.

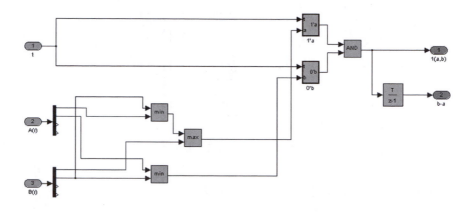

FIGURE 6.12C
The structure of the I2 unit.

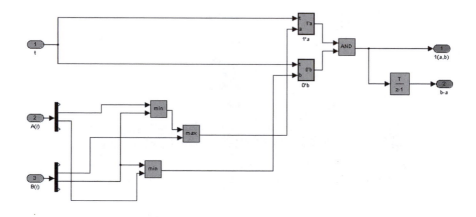

FIGURE 6.12D
The structure of the I3 unit.

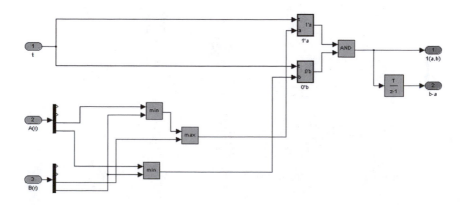

FIGURE 6.12E
The structure of the I4 unit.

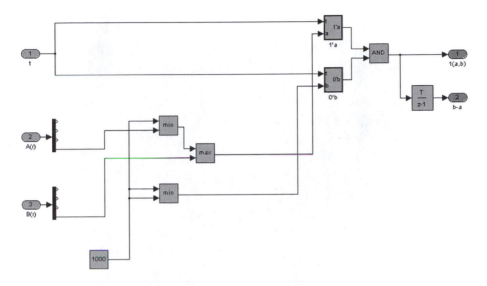

FIGURE 6.12F
The structure of the I5 unit.

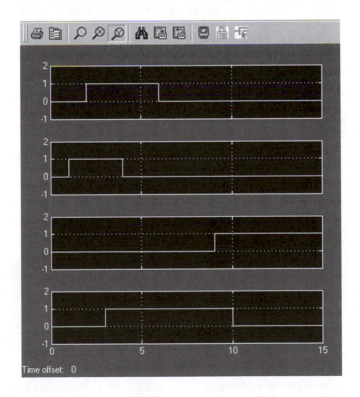

FIGURE 6.12G
Input impacts for the f_4^0 (t) implementing gate.

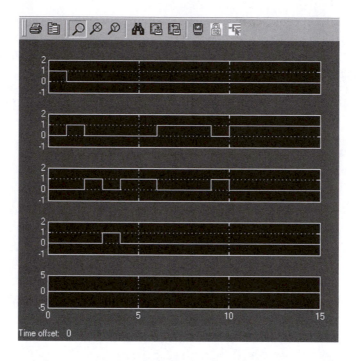

FIGURE 6.12H

Output impacts for the gates that implement the fundamental symmetrical functions of f_4^0, f_4^1, f_4^2, f_4^3, and f_4^4, respectively.

$a_{i,min}$ element should be determined for each set, and a maximal element should be chosen of $a_{i,min}$.

Using dedicated equivalent transformations, the a^r search algorithm of lower complexity can be obtained. The totality of search algorithms for all $LD - a^r$, $r = 1, 2,..., M = \Sigma m_i$ ($1 \leq i \leq n$) is the array sorting algorithm:

$$\{a_{11}, a_{12},..., a_{1m_1}, a_{21}, a_{22},..., a_{nm_n}\}.$$

The LD aids are closely related to the sorting networks. The sorting algorithm's essential features are the requirements for the algorithm's supplementary storage volume, in addition to the storage volume directly occupied by the initial sorting array and the sorting program.

Hardware application of the sorting algorithm is made possible by a dedicated device (sorting network), whose inputs are connected to all disordered array elements, while its outputs develop the ordered array. The hardware application of the sorting algorithm increases its speed, excluding a number of utility operations out of the sorting program, and overlapping a great number of different comparisons in time. This process uses no supplementary memory. The class of sorting algorithms under examination demands that the following condition be met (lack of memory): the sequence of comparisons of the array elements should be independent of its background. Because of that, some branches of the comparison tree need more than the otherwise necessary comparisons for the array ordering.

From the viewpoint of the LD facilities, the memory-free synthesis of sorting algorithms is mathematically reduced to the optimization of combined calculations of the Equation 6.47 search functions, defined at a finite set of combinations of discrete variables. The combinatorial programming used in this work is characterized by the extensive application of exhaustion techniques for the problem solutions. However, rapid growth of the calculation's complexity as the problem dimensions increase and the impossibility of visible presentation of the optimization algorithm for high-dimensional tasks send us in search of ways to improve the method.

A feasible approach is the structural representation (as the combination of interrelated parts) of sorted combinations of variables and exclusion of common parts by the comparison of both combinations. In this process the severity of finding an optimal solution decreases drastically. Such an approach demands specialized mathematical means that permit structural representation of the solution search algorithm. Both IVL and LD can serve as such [3, 4].

The memory-free sorting algorithm is conveniently represented by a logical network with comparators as elements (Figure 6.13). The expression for the sorting network r-th output in the $\{a_1,..., a_n\}$ disordered array through all its n inputs is the r-th rank LD of the column quasi-matrix:

$$A_n = \left\| \begin{matrix} a_1 \\ \cdots \\ a_n \end{matrix} \right\|; \qquad (6.49)$$

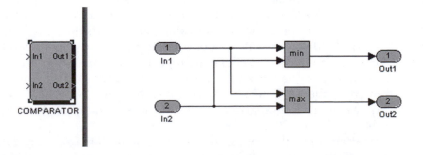

FIGURE 6.13
The structure of a comparator as the sorting network element.

If the array is partially ordered, A_n takes the form

$$A_n = \begin{Vmatrix} a_{11} \cdots a_{1m1} \\ \cdots\cdots\cdots \\ a_{n1} \cdots a_{nm1} \end{Vmatrix} ;$$

The totally ordered array is represented by line matrix

$$A_1 = \|a^1, a^2, \ldots, a^n\|.$$

The challenge is to find the multi-input sorting network of minimal cost (i.e., the network with the least comparators). First, we restrict our attention to the comparisons of neighboring elements and take into account that $n - 1$ comparisons are necessary and sufficient for finding the largest element of a given $\{a_1, \ldots, a_n\}$ array. After the necessary $n - 1$ comparisons of the array elements, the starting Equation 6.49 quasi-matrix looks like:

$$A'_n = \begin{Vmatrix} a_1 a_2 \\ (a_1 \vee a_2)a_3 \\ (a_1 \vee a_2 \vee a_3)a_4 \\ \cdots\cdots\cdots\cdots\cdots \\ (a_1 \vee a_2 \vee \ldots \vee a_{n-1})a_n \\ \hline a_1 \vee a_2 \vee \ldots \vee a_{n-1} \vee a_n \end{Vmatrix} \equiv \begin{Vmatrix} A'_{n-1} \\ \hline a^n \end{Vmatrix}$$

In this way the largest element of the $a_1 \vee a_2 \vee \ldots \vee a_{n-1} \vee a_n$ array equal to the A^n_n LD of the A_n quasi-matrix has taken its proper place at the array end (we will consider it as the sorted $\{A^n_n\}$ subarray of the A_n array). Then, using the $(n - 1) - 1$ comparisons, we calculate A^{n-1}_n (isolate the largest element out of the disordered A'_{n-1} subarray) and thus increase to two the number of the sorted subarray elements, repeating the operation until the entire A_n source array is sorted out. Such an algorithm of the joint LD calculation from the $\{A^1_n, A^2_n, \ldots, A^n_n\}$ family has the $\Sigma(i - 1) = O(n^2)$ complexity of $(1 \le i \le n)$ comparisons and is the so-called bubble technique (Figure 6.14).

An alternative algorithm of the joint LD calculation from the $\{A^1_n, A^2_n, \ldots, A^n_n\}$ family is described by the following theorem: the ratio of the A^r_m ($r = 1, 2, \ldots, m$) LD for the matrix and the A^r_{m+1} determinant for the A_{m+1} matrix obtained from A_m by inclusion of an additional a_{m+1} element

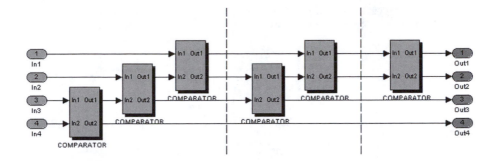

FIGURE 6.14
The sorting network for the four-number array, the bubble technique (method).

is determined in terms of IVL conjunctions and disjunctions by the following expression:

$$A^r_{m+1} = \begin{cases} A^r_m\, a_{m+1}, & r=1; \\ A^r_m\, a_{m+1} \vee A^{r-1}_m, & r=2,3,\dots m; \\ A^{r-1}_m \vee a_{m+1}, & r=m+1. \end{cases} \qquad (6.50)$$

Let us isolate as the source, the basic element of any arbitrary element of the disordered array. It will form the sorted subarray $\{A^1_1\}$ (made of one element) of the array. Using the expressions of Equation 6.50, we can add an additional (neighboring) element to this subarray. As a result, we have a sorted $\{A^1_2, A^2_2\}$ subarray and proceed with the process until the entire array is sorted, that is, until it takes the form of $\{A^1_n, A^2_n,\dots, A^n_n\}$. Such is the insertion technique, and its complexity has $O(n^2)$ comparisons (Figure 6.15). The bubble technique and the insertion technique are reduced to the canonical "triangle" procedure common for both (Figure 6.16).

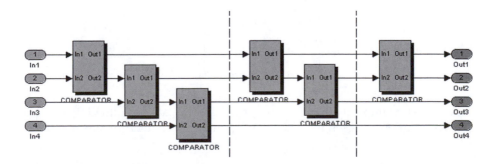

FIGURE 6.15
The sorting network for the four-number array: the insertion technique (method).

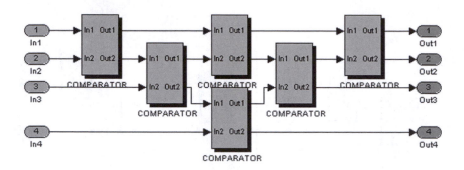

FIGURE 6.16
The sorting network for the four-number array: the triangle technique (method).

If we do not restrict ourselves by the comparisons of neighboring elements only, we can obtain the lower complexity algorithms for joint computation of the LD family — up to the $O(n(\log n)^2)$ comparisons. A general rule of LD exposure is possible that consists in its decomposition-based gradual decrease in order:

$$A^r_n = \left| \begin{array}{c} A_{n \setminus i1,\ldots, ik} \\ A^1_{i1,\ldots, ik} \ldots A^P_{i1,\ldots, ik} \end{array} \right|^r , \quad \begin{array}{l} p = \Sigma m_{is}; \\ (1 \le s \le k) \end{array} \tag{6.51}$$

where: $A_{n \setminus i1,\ldots,ik}$ is the A_n quasi-matrix with removed $i1,\ldots,$ ik rows; $A^P_{i1,\ldots,ik}$ is the p-th rank k-th order determinant composed of $i1,\ldots,ik$ rows until the expression that contains only second-order determinants is obtained.

An effective algorithm for joint calculation of the $\{A^1_2, A^2_2,\ldots, A^r_2,\ldots, A^{m1+m2}_2\}$, $m1 \ge m2$ family logical determinants should be obtained, using:

$$A^r_2 = \begin{cases} \max a_{1i} a_{2,r+1-i}, & ((1 \le i \le r), \quad r = 1,\ldots, m2); \\ a_{1,r-m2} \vee \max a_{1i}a_{2,r+1-i}, & ((r \le i \le r - m2 + 1), r = m2 + 1,\ldots, m1); \\ a_{1,r-m2} \vee \max a_{1i}a_{2,r+1-i}, & ((m1 \le i \le r - m2 + 1), \quad r = m1 + 1,\ldots, m1 + m2). \end{cases} \tag{6.52}$$

Assume for simplicity that $m1 = m2 = m$ and formulate the following theorem [5]: The relation between the A^r_2 $(r = 1,\ldots, 2m)$ logical determinants for the

$$A_2 = \left\| \begin{array}{c} a_{11} a_{12} \ldots a_{1m} \\ a_{21} a_{22} \ldots a_{2m} \end{array} \right\|$$

matrix and for the

$$K_2^r \ (r = 1, 2, \ldots, 2(2k + 1) \le 2m,$$

$$L_2^r \ (r = 1, 2, \ldots, 2(2l) \le 2m$$

determinants of its submatrices

$$K_2 = \begin{Vmatrix} a_{11}a_{13}\ldots a_{1,2k+1} \\ a_{21}a_{23}\ldots a_{2,2k+1} \end{Vmatrix}$$

$$L_2 = \begin{Vmatrix} a_{12}a_{14}\ldots a_{1,2l} \\ a_{22}a_{24}\ldots a_{2,2l} \end{Vmatrix}$$

(K_2 contains all odd columns of A_2, while L_2 contains all even columns of A_2) is determined in terms of IVL disjunctions and conjunctions by

$$A^1_2 = K^1_2$$

$$\ldots\ldots\ldots\ldots\ldots$$

$$A^{2i}_2 = L^i_2 K^{i+1}_2$$

$$A^{2i+1}_2 = L^i_2 \vee K^{i+1}_2 \tag{6.53}$$

$$\ldots\ldots\ldots\ldots$$

$$A^{2m}_2 = L^m_2, \text{ m is even;}$$

$$K^{m+1}_2, \text{ m is odd.}$$

The proof of the Equation 6.53 relation is obtained by m induction. With m = 1, Equation 6.53 takes the form of evident relation:

$$A_2 = K_2 = \begin{Vmatrix} a_{11} \\ a_{21} \end{Vmatrix}, A^1_2 = a_{11}a_{21} = K^1_2, A^2_2 = a_{11} \vee a_{21} = K^2_2,$$

With m = 2, it is also evident that

$$A_2 = \begin{Vmatrix} a_{11} & a_{12} \\ a_2 & a_{22} \end{Vmatrix}, K_2 = \begin{Vmatrix} a_{11} \\ a_{21} \end{Vmatrix}, L_2 = \begin{Vmatrix} a_{12} \\ a_{22} \end{Vmatrix}.$$

$$A^1_2 = a_{11} a_{21} = K^1_2, A^2_2 = a_{11} a_{22} \vee a_{12}a_{21} = a_{12} a_{22}(a_{11} \vee a_{21}) = L^1_2K^2_2,$$

$$A^3_2 = (a_{11} \vee a_{21}) \vee a_{12} a_{22} = L^1_2 \vee K^2_2, A^4_2 = a_{12} \vee a_{22} = L^2_2.$$

Assume that Equation 6.52 is valid for a certain $m = p$. We can show that in this case it is also valid for $m = p + 1$. Let m be even. Then, Equation 6.52 is valid for even m. Now we should order the attached $(p + 1)$-th column:

$$
\left| \begin{matrix} a_{1,p+1} \\ a_{2,p+1} \end{matrix} \right|^r = \begin{cases} a_{1,p+1}\, a_{2,p+1}, & r = 1; \\ a_{1,p+1} \vee a_{2,p+1}, & r = 2. \end{cases}
$$

It is evident that

$$
A^r_2 = \left| \begin{matrix} a_{11}\cdots a_{1p}a_{1,p+1} \\ a_{21}\cdots a_{2p}a_{2,p+1} \end{matrix} \right|^r
$$

$$
= \left| \begin{matrix} K^1_2\left(L^1_2 K^2_2\right)\left(L^1_2 \vee K^2_2\right)\cdots\left(L^i_2 K^{i+1}_2\right)\left(L^i_2 \vee K^{i+1}_2\right)\cdots\left(L^{p-1}_2 K^p_2\right)\left(L^{p-1}_2 \vee K^p_2\right) \\ L^p_2 a_{1,p+1}\, a_{2,p+1} \quad a_{1,p+1} \vee a_{2,p+1} \end{matrix} \right|^r
$$

$$(6.54)$$

According to Equation 6.52 (with $m_1 = 2p$, $m_2 = 2$), exposure of Equation 6.53 looks like

$$
A^r_2 = \begin{cases}
K^1_2(a_{1,p+1}a_{2,p+1}), & r = 1; \\[6pt]
K^1_2(a_{1,p+1} \vee a_{2,p+1}) \vee \left(L^1_2 K^2_2\right)(a_{1,p+1}a_{2,p+1}), & r = 2; \\[6pt]
\left(L^1_2 K^2_2\right)(a_{1,p+1} \vee a_{2,p+1}) \vee \left(L^1_2 \vee K^2_2\right)(a_{1,p+1}a_{2,p+1}) \vee K^1_2, & r = 3; \\[6pt]
\left(L^1_2 \vee K^2_2\right)(a_{1,p+1} \vee a_{2,p+1}) \vee \left(L^1_2 K^3_2\right)(a_{1,p+1}a_{2,p+1}) \vee L^1_2 K^2_2, & r = 4; \\[6pt]
\left(L^1_2 \vee K^3_2\right)(a_{1,p+1} \vee a_{2,p+1}) \vee \left(L^2_2 \vee K^3_2\right)(a_{1,p+1}a_{2,p+1}) \vee \left(L^1_2 \vee K^2_2\right), & r = 5; \\[6pt]
\hdashline \\[-4pt]
L^p_2(a_{1,p+1} \vee a_{2,p+1}) \vee \left(L^{p-1}_2 \vee K^p_2\right)(a_{1,p+1}a_{2,p+1}), & r = 2p+1; \\[6pt]
L^p_2 \vee (a_{1,p+1} \vee a_{2,p+1}), & r = 2(p+1).
\end{cases}
$$

$$(6.55)$$

Take into account the addition of two $(p + 1)$-th column elements to the K_2 quasi-matrix (L_2 is invariable, and K_2 converts to the K_2' quasi-matrix).

$$(K_2') = \begin{cases} K^1_2(a_{1,p+1}a_{2,p+1}), & r = 1; \\[6pt] K^1_2(a_{1,p+1} \vee a_{2,p+1}) \vee K^2_2(a_{1,p+1}a_{2,p+1}), & r = 2; \\[6pt] K^2_2(a_{1,p+1} \vee a_{2,p+1}) \vee K^3_2(a_{1,p+1}a_{2,p+1}) \vee K^1_2, & r = 3; \\[6pt] K^3_2(a_{1,p+1} \vee a_{2,p+1}) \vee K^4_2(a_{1,p+1}a_{2,p+1}) \vee K^2_2, & r = 4; \\[6pt] \text{\textemdash\textemdash\textemdash\textemdash\textemdash\textemdash\textemdash\textemdash\textemdash\textemdash\textemdash\textemdash} & \\[6pt] K^P_2(a_{1,p+1} \vee a_{2,p+1}) \vee K^{P-1}_2 \vee (a_{1,p+1}a_{2,p+1}), & r = p+1; \\[6pt] K^P_2 \vee (a_{1,p+1} \vee a_{2,p+1}), & r = 2(p+1). \end{cases}$$ (6.56)

Having performed some simple transformations (using the IVL laws) in Equation 6.55 and having taken the Equation 6.56 relation into account, we have:

$$A^r_2 = \begin{cases} (K_2')^1, & r = 1; \\[6pt] L^1_2\left[K^1_2(a_{1,p+1} \vee a_{2,p+1}) \vee K^2_2(a_{1,p+1}a_{2,p+1})\right] = L^1_2(K_2')^1, & r = 2; \\[6pt] L^1_2 \vee \left[K^1_2(a_{1,p+1} \vee a_{2,p+1}) \vee K^2_2(a_{1,p+1}a_{2,p+1})\right] = L^1_2 \vee (K_2')^1, & r = 3; \\[6pt] L^2_2\left[K^2_2(a_{1,p+1} \vee a_{2,p+1}) \vee K^3_2(a_{1,p+1}a_{2,p+1}) \vee K^1_2\right] = L^2_2(K_2')^3, & r = 4; \\[6pt] L^2_2 \vee \left[K^2_2(a_{1,p+1} \vee a_{2,p+1}) \vee K^3_2(a_{1,p+1}a_{2,p+1}) \vee K^1_2\right] = L^2_2 \vee (K_2')^3, & r = 5; \\[6pt] \text{\textemdash\textemdash\textemdash\textemdash\textemdash\textemdash\textemdash\textemdash\textemdash\textemdash\textemdash} & \\[6pt] L^P_2 \vee \left[K^P_2(a_{1,p+1} \vee a_{2,p+1}) \vee K^{P-1}_2 \vee (a_{1,p+1}a_{2,p+1})\right] = L^P_2 \vee (K_2')^{P+1}, & r = 2p+1; \\[6pt] L^P_2 \vee (a_{1,p+1} \vee a_{2,p+1}) = (K_2')^{(p+1)}, & r = 2(p+1). \end{cases}$$

Thus we have obtained the Equation 6.53 relation with odd $m = p + 1$, that is, the Equation 6.53 relation, if it holds for even $m = p$, remains valid for $m = p+1$ as well. For odd m the proof is similar. The theorem has been proved.

The sorting algorithm described above, using the Equation 6.53 relation, was invented by Batcher. (Batcher's method is described in reference 8.) Its complexity — the number of sorting comparators for the disordered array of n numbers — is $0(n[\log_2 n]^2)$. The sorting time (in comparator delay units) is $0(\log_2 n)$ (i.e., most of comparisons are parallel).

6.3 Model Building for Timing Verification

6.3.1 Spectrum Analysis of a Switching Process

The model for spectrum analysis of switching processes is shown in Figure 6.17. Such analysis is performed by calculating the fundamental f_n^i, $i = 0, 1,..., n$ symmetric functions [5–7]. Each fundamental symmetric function, differing from zero at a certain time interval, represents a *spectral line* within the switching process spectrum. The f_n^i is calculated in the i index growth sequence. Note that it is not obligatory to calculate all $f_n^i : f_n^0, f_n^1,...,$ f_n^n for each specific case. Spectrum analysis of a switching process is over as soon as the equality

$$\Sigma \, T(f_n^{\ i}) = \max b_j, \ (j = 1,..., n; \ i = 0,..., k) \tag{6.57}$$

holds. Here $T(f_n^i)$ is a net duration of the f_n^i (t) switching process pulses, n is the number of combinational circuit inputs (the switching process dimensionality), $k \leq n$, and max b_j is the upper limit of time interval for the analyzed switching process.

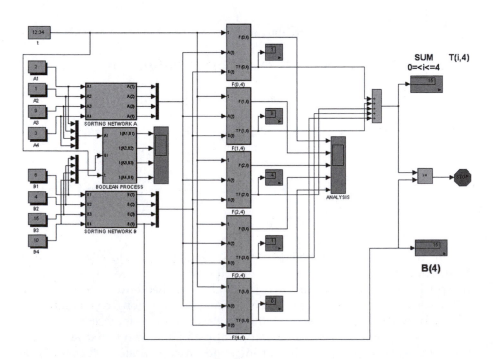

FIGURE 6.17
The model for the switching process of spectrum analysis.

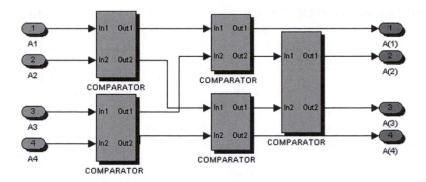

FIGURE 6.18
The sorting network for the pulse starting points in the analyzed switching process.

In Figure 6.17, the input data sources are the pulse beginning and end points for the analyzed switching process A1, A2, A3, A4, B1, B2, B3, and B4 (in our specific case, the number of combinational circuit inputs n = 4, and only one pulse is fed for each input [cf. Figure 6.43]). The Digital Clock Block t sets the modeling time.

Seven Display Blocks indicate the net pulse lengths for the $T(f_n^i)$, (i = 0, 1, 2, 3, 4) switching processes as well as the $\Sigma T(f_n^i)$ sum and the upper limit of the time interval of the analyzed switching process, B(4). Two Scope Blocks permit viewing of the analyzed switching process (Figure 6.43), and the spectrum analysis results (Figure 6.44).

The Stop Simulation Block is red, and it stops the analysis when a pulse is received from the Relational Operator ($\geq$) Block at the point when the Equation 6.56 equality is achieved. The rest of model units (cyan) are intended for information processing. The SORTING NETWORK A and SORTING NETWORK B units have similar structure shown in Figure 6.18. The units sort the beginning and end points, respectively, of the analyzed switching process pulses. The units contain five comparators each, and their inner structure is shown in Figure 6.19.

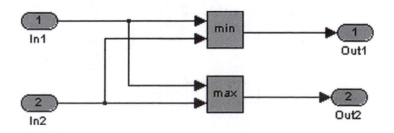

FIGURE 6.19
The comparator structure.

The BOOLEAN PROCESS unit generates the switching process from the known points of pulse beginnings and ends. Four pulses, 1(A1,B1), 1(A2,B2), 1(A3,B3), and 1(A4,B4), are generated at the unit output (see Figure 6.20). Each pulse is generated by means of three units: $1'_a$, $0'_b$, and AND. Two Mux Blocks ensure the delivery of the pulse beginning and end points to all $1'_a$ and $0'_b$ units. Internal structures of the $1'_a$ and $0'_b$ units are shown in Figure 6.21 and Figure 6.22.

The F(0,4), F(1,4), F(2,4), F(3,4), and F(4,4) units are tailored for the calculation of fundamental symmetrical functions of f_n^i $i = 0, 1, 2, 3, 4$. Each such

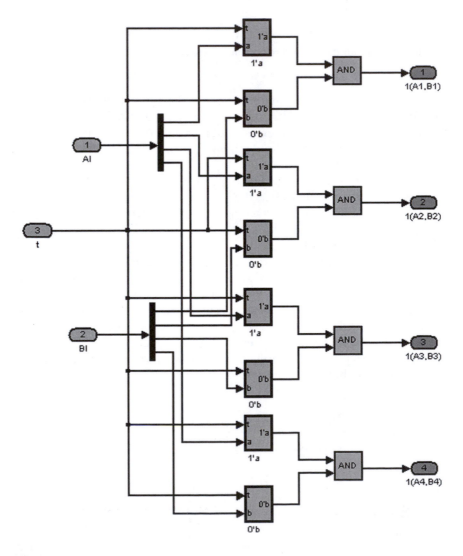

FIGURE 6.20
The BOOLEAN PROCESS unit structure.

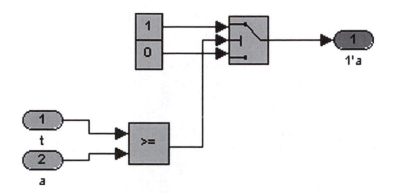

FIGURE 6.21
The $1'_a$ unit structure incorporated into the BOOLEAN PROCESS unit.

unit has three inputs: t, A(r), and B(r). The first input is the synchronization input, corresponding to the simulation time course. Two other inputs receive the data on pulse beginnings and ends for the analyzed switching process.

Each of these units has two outputs: F(i,4) and TF(i,4), where i = 0, 1, 2, 3, 4. At the first output a switching process is generated that is described by the corresponding fundamental symmetrical function of f_n^i. The second unit determines the switching process's net duration — $T(f_n^i)$. Internal structures of the F(0,4), F(1,4), F(2,4), F(3,4), and F(4,4) units are shown in Figures 6.23 to 6.27.

Each F(i,4) unit incorporates one to five In units that generate certain pulses of the corresponding fundamental symmetrical function $f_4^i(t)$. The F(0,4) unit

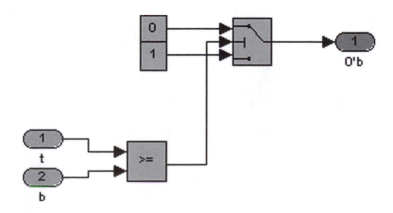

FIGURE 6.22
The $0'_b$ unit structure incorporated into the BOOLEAN PROCESS unit.

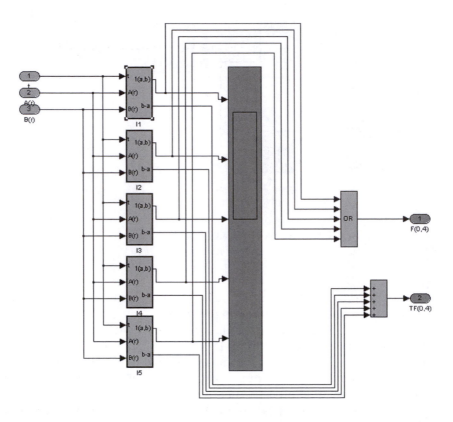

FIGURE 6.23
The structure of the F(0,4) unit that implements the fundamental symmetrical function, f_4^0 (t) $= \Sigma 1(b^{j-1}, a^j)$, $(1 \le j \le 5)$.

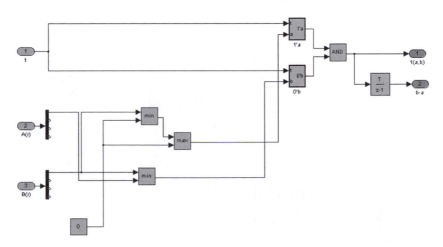

FIGURE 6.24
The structure of the F(1,4) unit that implements the fundamental symmetrical function, f_4^1 (t) $= \Sigma 1(b^{j-1} \vee a^j, b^j a^{j+1})$, $(1 \le j \le 4)$.

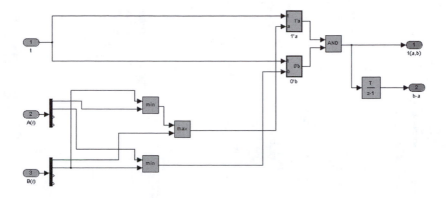

FIGURE 6.25
The structure of the F(2,4) unit that implements the fundamental symmetrical function, f_4^2 (t) $= \Sigma 1(b^{j-1} \vee b^j a^{j+1}, b^j a^{j+2})$, $(1 \leq j \leq 3)$.

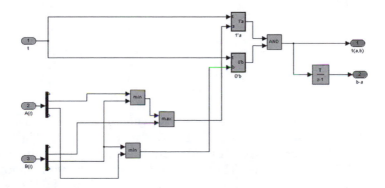

FIGURE 6.26
The structure of the F(3,4) unit that implements the fundamental symmetrical function, f_4^3 (t) $= \Sigma 1(b^{j-1} \vee b^j a^{j+2}, b^j a^{j+3})$, $(1 \leq j \leq 2)$.

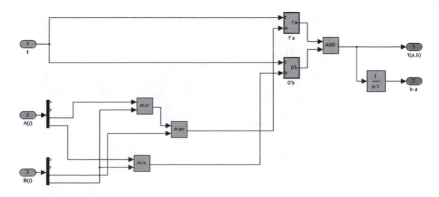

FIGURE 6.27
The structure of the F(4,4) unit that implements the fundamental symmetrical function, f_4^4 (t) $= 1(b^{j-1} \vee b^j a^{j+3}, b^j a^{j+4})$, $(j = 1)$.

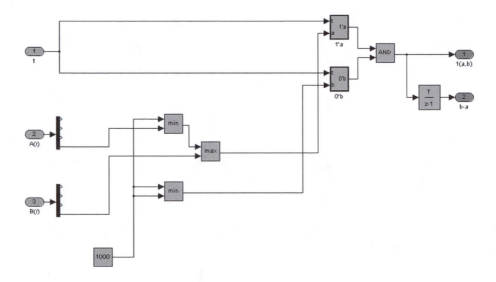

FIGURE 6.28
The structure of the I1 unit that enters into the F(0,4) unit and generates the $1(b^{j-1}, a^j) \equiv 1(0, a^1)$, $j = 1$ pulse.

has five such units, and their structures are given inFigures 6.28 to 6.32. The F(1,4) unit has four such units, and their structures are given in Figures 6.33 to 3.36. The F(2,4) unit has three In units, and their structures can be found in Figures 6.37 to 6.39. The F(3,4) unit contains two In units, and their structures can be found in Figures 6.40 and 6.41. Finally, the F(4,4) unit contains only one I1 unit, and its structure is shown in Figure 6.42. In addition, each F(i,4) unit contains the OR unit for logical summation of said pulses and an adder for the calculation of their net duration for each $f^i_4(t)$ switching process.

Recall that in our particular case $n = M = 4$.

$$f_4^0 (t) = \Sigma 1(b^{j-1}, a^j), (1 \leq j \leq 5);$$

$$f_4^1 (t) = \Sigma 1(b^{j-1} \vee a^j, b^j a^{j+1}), (1 \leq j \leq 4);$$

$$f_4^2 (t) = \Sigma 1(b^{j-1} \vee b^j a^{j+1}, b^j a^{j+2}), (1 \leq j \leq 3); \qquad (6.58)$$

$$f_4^3 (t) = \Sigma 1(b^{j-1} \vee b^j a^{j+2}, b^j a^{j+3}), (1 \leq j \leq 2);$$

$$f_4^4 (t) = \Sigma 1(b^{j-1} \vee b^j a^{j+3}, b^j a^{j+4}), (1 \leq j \leq 1).$$

Inside each In unit a Discrete-Time Integrator Block is used for the calculation of b-a duration generated by the In unit of the 1(a, b) pulse.

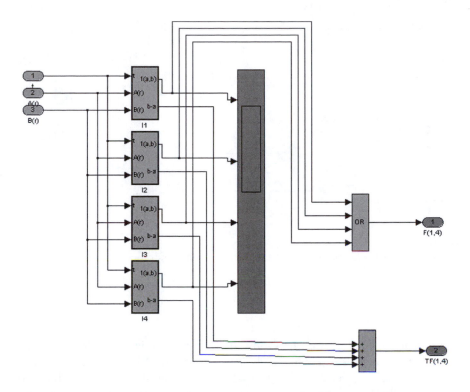

FIGURE 6.29
The structure of the I2 unit that enters into the F(0,4) unit and generates the $1(b^{i-1}, a^i) \equiv 1(b^1, a^2)$, $j = 2$ pulse.

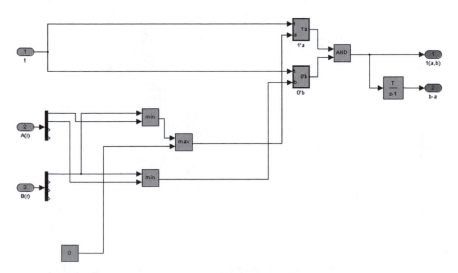

FIGURE 6.30
The structure of the I3 unit that enters into the F(0,4) unit and generates the $1(b^{i-1}, a^i) \equiv 1(b^2, a^3)$, $j = 3$ pulse.

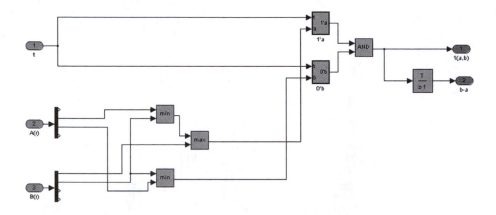

FIGURE 6.31
The structure of the I4 unit that enters into the F(0,4) unit and generates the $1(b^{j-1}, a^j) \equiv 1(b^3, a^4)$, $j = 4$ pulse.

As mentioned before, the analyzed switching process is shown in Figure 6.43. It contains four pulses: 1(2,6), 1(1,4), 1(9,15), and 1(3,10). The spectrum analysis results for a given switching process are shown in Figure 6.44. It can be seen from the figure that the spectrum of the analyzed switching process

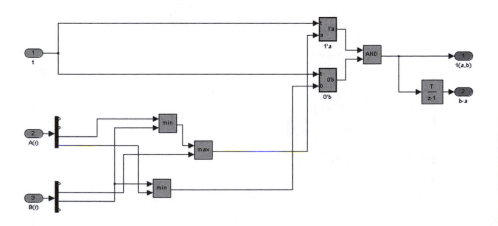

FIGURE 6.32
The structure of the I5 unit that enters into the F(0,4) unit and generates the $1(b^{j-1}, a^j) \equiv 1(b^4, a^5)$, $j = 5$ pulse.

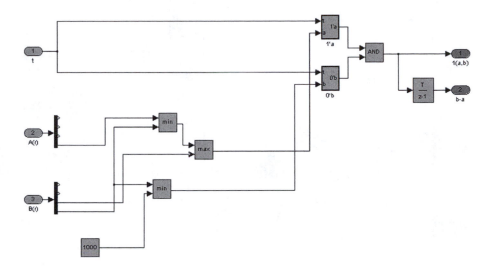

FIGURE 6.33
The structure of the I1 unit that enters into the F(1,4) unit and generates the $1(b^{j-1} \lor a^j, b^j a^{j+1}) \equiv 1(a^1, b^1 a^2), j = 1$ pulse.

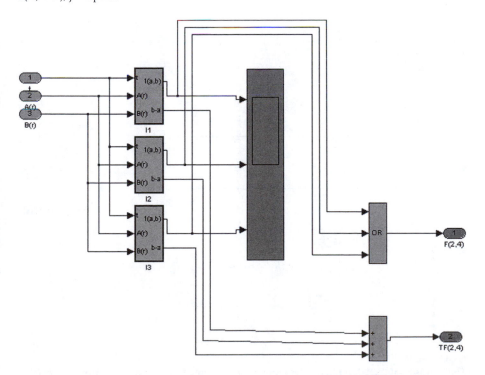

FIGURE 6.34
The structure of the I2 unit that enters into the F(1,4) unit and generates the $1(b^{j-1} \lor a^j, b^j a^{j+1}) \equiv 1(b^1 \lor a^2, b^2 a^3), j = 2$ pulse.

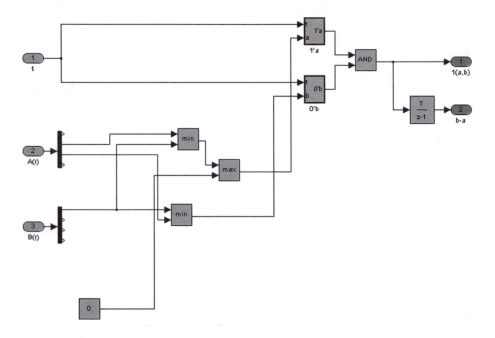

FIGURE 6.35
The structure of the I2 unit that enters into the F(1,4) unit and generates the $1(b^{j-1} \vee a^j, b^j a^{j+1}) \equiv 1(b^2 \vee a^3, b^3 a^4)$, $j = 3$ pulse.

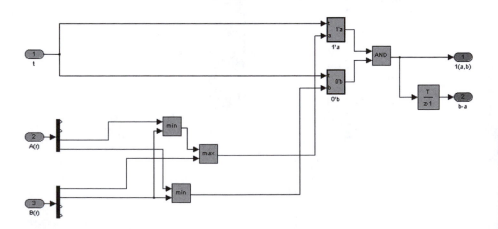

FIGURE 6.36
The structure of the I4 unit that enters into the F(1,4) unit and generates the $1(b^{j-1} \vee a^j, b^j a^{j+1}) \equiv 1(b^3 \vee a^4, b^4 a^5)$, $j = 4$ pulse.

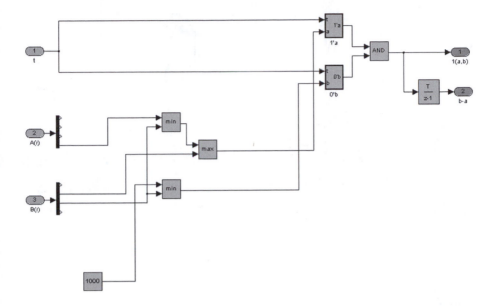

FIGURE 6.37
The structure of the I1 unit that enters into the F(2,4) unit and generates the $1(b^{j-1} \vee b^j a^{j+1}, b^j a^{j+2})$ $\equiv 1(b^1 a^2, b^1 a^3)$, j = 1 pulse.

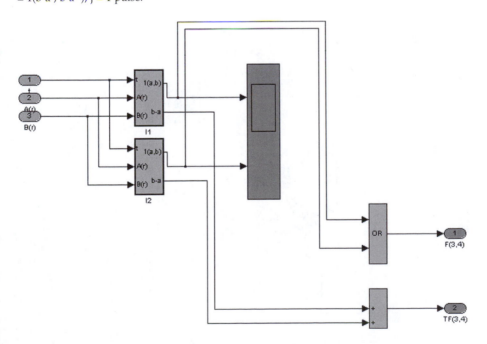

FIGURE 6.38
The structure of the I2 unit that enters into the F(2,4) unit and generates the $1(b^{j-1} \vee b^j a^{j+1}, b^j a^{j+2})$ $\equiv 1(b^1 \vee b^2 a^3, b^2 a^4)$, j = 2 pulse.

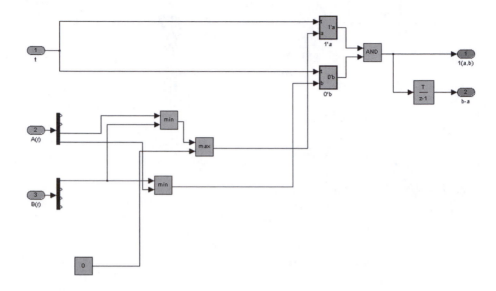

FIGURE 6.39
The structure of the I3 unit that enters into the F(2,4) unit and generates the $1(b^{j-1} \vee b^j a^{j+1}, b^j a^{j+2})$ $\equiv 1(b^2 \vee b^3 a^4, b^3 a^5)$, j = 3 pulse.

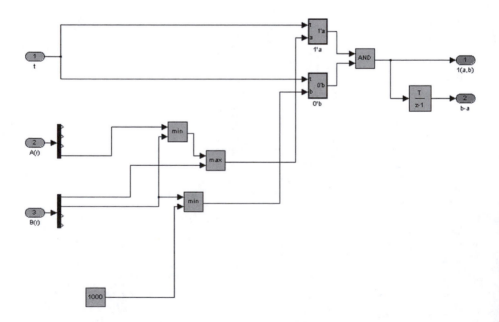

FIGURE 6.40
The structure of the I1 unit that enters into the F(3,4) unit and generates the $1(b^{j-1} \vee b^j a^{j+2}, b^j a^{j+3})$ $\equiv 1(b^1 a^3, b^1 a^4)$, j = 1 pulse.

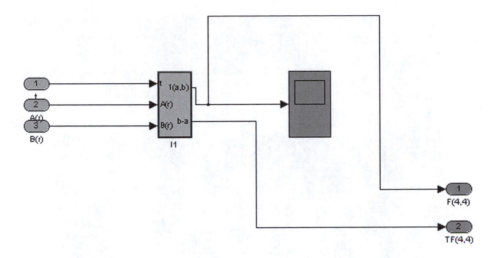

FIGURE 6.41

The structure of the I2 unit that enters into the F(3,4) unit and generates the $1(b^1 \vee b^2a^4, b^2a^5) \equiv 1(b^1a^3, b^1a^4)$, j = 2 pulse.

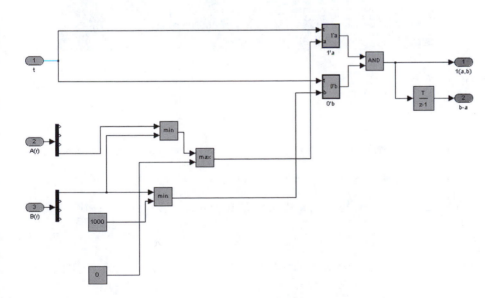

FIGURE 6.42

The structure of the I1 unit that enters into the F(4,4) unit and generates the $1(b^{j-1} \vee b^ja^{j+3}, b^ja^{j+4}) \equiv 1(b^1a^4, b^1)$, j = 1 pulse.

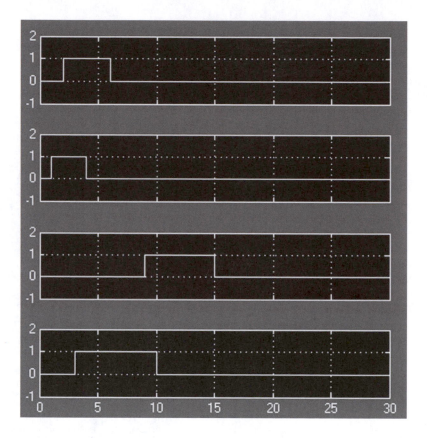

FIGURE 6.43
The analyzed switching process.

has four nonzero spectral lines (the calculations are based on Equation 6.58):

f_4^0 (t) = $\Sigma 1(b^{j-1}, a^j)$, $(1 \le j \le 5) \Rightarrow \Rightarrow f_4^0$ (t) = 1(0,1);

f_4^1 (t) = $\Sigma 1(b^{j-1} \vee a^j, b^j a^{j+1})$, $(1 \le j \le 4) \Rightarrow \Rightarrow f_4^1$ (t) = 1(1,2)0(–,–)1(6,9) 0(–,–)1(10,15);

f_4^2 (t) = $\Sigma 1(b^{j-1} \vee b^j a^{j+1}, b^j a^{j+2})$, $(1 \le j \le 3) \Rightarrow \Rightarrow f_4^2$ (t) = 1(2,3)0(–,–)1(4,6) 0(–,–)1(9,10);

f_4^3 (t) = $\Sigma 1(b^{j-1} \vee b^j a^{j+2}, b^j a^{j+3})$, $(1 \le j \le 2) \Rightarrow \Rightarrow f_4^3$ (t) = 1(3,4);

f_4^4 (t) = $\Sigma 1(b^{j-1} \vee b^j a^{j+3}, b^j a^{j+4})$, $(1 \le j \le 1) \Rightarrow \Rightarrow f_4^4$ (t) = 0.

This is the qualitative portrait of the analyzed switching process, which means that with minor displacements of the pulse beginnings and ends with respect to their initial condition, the process spectrum will contain the same spectral lines.

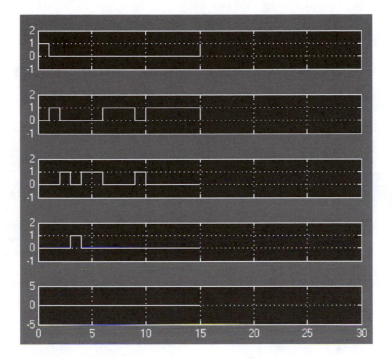

FIGURE 6.44
The spectrum analysis results for the switching process.

6.3.2 The Comparison Model for Two Switching Processes

The comparison model for two switching processes is shown in Figure 6.45. Again, the Digital Clock Block t sets the model time. The data on model time progress arrives at the inputs of BOOLEAN PROCESS 1 and BOOLEAN PROCESS 2 units that generate the switching processes to be compared and to the inputs of F(0,2), F(1,2), and F(2,2) units that calculate the fundamental symmetrical functions f_2^0 (t), f_2^1 (t), and f_2^2 (t), respectively. The three-input Sum Block provides the calculation of net duration of pulses for the above-mentioned fundamental symmetrical functions, $\sum T(f_2^{\ i})$, (i = 0,1,2). The Selector Block makes it possible to obtain B(4), the upper limit of the time interval for the compared switching processes (in our particular case both switching processes contain only two pulses each). Two SORTING NETWORK A, SORTING NETWORK B units sort the pulse beginning (A_{11}, A_{12}, A_{21}, A_{22}) and end points (B_{11}, B_{12}, B_{21}, B_{22}) of the compared switching processes, respectively (Figure 6.48).

The Stop Simulation Block stops the analysis when a pulse is received from the Relational Operator ($\geq$) Block at the point when the Equation 6.56 equality is achieved.

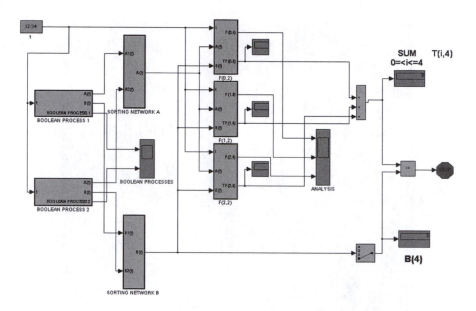

FIGURE 6.45
The comparison model for two switching processes.

Five Display Blocks indicate the net pulse lengths for the $T(f_2^i)$, ($i = 0, 1, 2$), switching processes as well as the $\sum T(f_2^i)$ sum and the upper limit of the time interval of the analyzed switching process, $B(4)$. Two Scope Blocks permit viewing of the compared switching processes (Figures 6.49 and 6.50) and their comparison results (Figures 6.51 and 6.52).

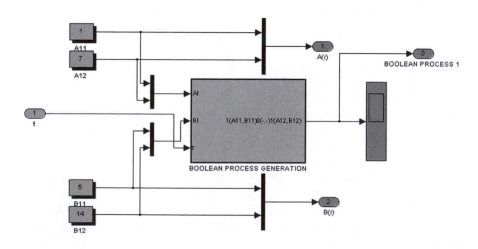

FIGURE 6.46
The BOOLEAN PROCESS 1,2 unit structures.

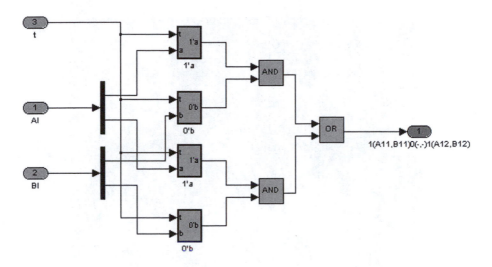

FIGURE 6.47
The BOOLEAN PROCESS GENERATION unit structure.

The BOOLEAN PROCESS 1,2 units have the same structure represented in Figure 6.46. Each such unit contains the Constant Blocks — to assign the pulse beginning and end points for a given switching process. As already mentioned, each process of our example has two pulses:

BOOLEAN PROCESS 1: A11=1, A12=7, B11=5, B12=14
BOOLEAN PROCESS 2: A21=2, A22=8, B21=6, B22=15

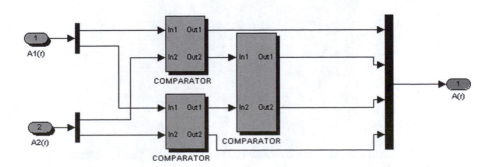

FIGURE 6.48
The SORTING NETWORK unit structure.

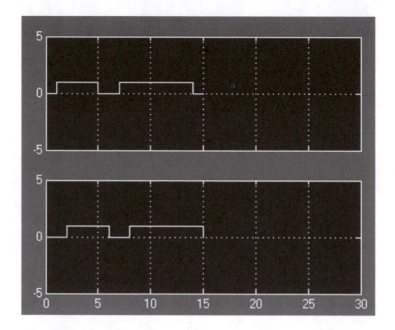

FIGURE 6.49
The switching processes to be compared (version 1).

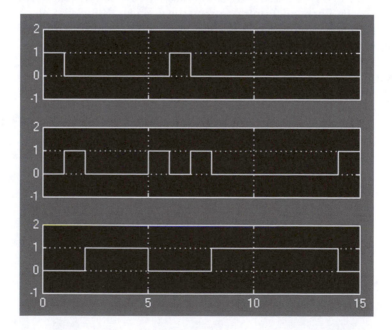

FIGURE 6.50
The switching processes to be compared (version 2).

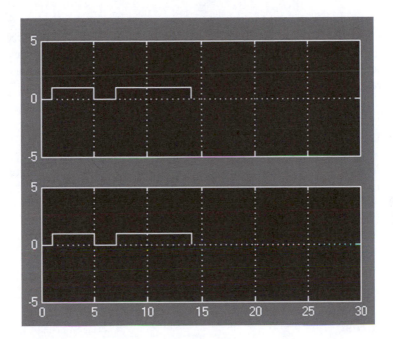

FIGURE 6.51
The switching processes' comparison results (version 1).

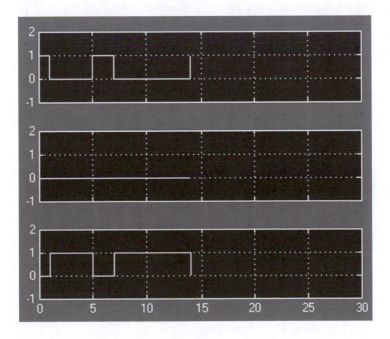

FIGURE 6.52
The switching processes' comparison results (version 2).

The multiplexors found inside the units form the vectors that contain the pulse beginnings and ends. The BOOLEAN PROCESS GENERATION units (their structures are given in Figure 6.47), based on the input data on the pulse beginning and end points, generate two switching processes to be compared:

BOOLEAN PROCESS 1: 1(1,5)0(−,−)1(7,14)

BOOLEAN PROCESS 2: 1(2,6)0(−,−)1(8,15) (cf. Figure 6.49)

The comparison results for these switching processes are shown in Figure 6.51. They bear witness to their inequality. When two switching processes are similar (Figure 6.50), the $f_2{}^1$ spectral line is lacking in the analysis results, and the following relation is found between the $f_2{}^0$ and $f_2{}^2$ spectral lines:

$$f_2{}^0 = \overline{f_2{}^2}.$$

References

1. Levin V.I. *Dynamics of Logical Units and Systems* (in Russian). Energia, Moscow, 1980.
2. Levin V.I. *Logical Reliability Theory of Complex Systems* (in Russian). Energoatomizdat, Moscow, 1985.
3. Levin V.I., Perelroyzen E.Z. Machine perception of spatial scenes. *Problems of Control and Information Theory*, 11(1), 53–66, 1982.
4. Levin V.I., Perelroyzen E.Z. Logical methods of spatial scenes researches. *Foundations of Control Engineering*, 7(4), 1982.
5. Perelroyzen E. Raster technology and parallel computations in simulation (in Russian). *Electronics Simulation*, 10(4), 1988.
6. Perelroyzen E. Automata Models for Space Scenes Researches (in Russian). PhD thesis., Riga Polytechnic Institute, Riga, Latvia, 1982.
7. Perelroyzen E. Self-learning in the task of 3d scenes analysis by dynamic planar projections using voronoi diagrams (in Russian). *Proceedings of Academy of Sciences of the USSR*, Series Control Engineering, #5, 1987.
8. Knuth D.E. *The Art of Computer Programming*, Vol. 3, *Sorting and Searching*. Addison-Wesley, Reading, MA, 1973.

7

System and Embedded Core Testing

7.1 Introduction

System tests can be subdivided into two large groups: functional tests and diagnostic tests. Functional tests verify system integrity using functional specifications. These are the tests of the pass or fail type with limited diagnostics capacities. Diagnostic tests are used after the system is found to have faults. Their objective is to locate the fault wherever they are located within the system. A diagnostic test employs the fault dictionary and the fault tree or diagnostic tree. The Fault dictionary contains a kit of test symptoms that can be associated with each simulated fault. The procedure called the Fault Tree uses the system of tests, and each of them is put to use at a particular level of the Fault Tree, helping to make a partial decision (diagnosis).

Making the partial decision chain eventually results in the final diagnosis that corresponds to a Fault Tree leaf. These tests are used together with scan path architectures and techniques, permitting the user to dissect the system into simpler constituents. As the synthesis of the deterministic test sequences as the input stimuli (for instance, by the automatic test pattern generation [ATPG] techniques) is quite complicated, the pseudo random sequences are used, coupled with built-in self-test (BIST) architecture) [1,2].

7.2 Scan Path Architectures and Techniques

7.2.1 Models for BIST Architecture

Test systems comprise a tested object (design-under-test [DUT]) subsystem for test access generation and for the analysis of DUT responses to these stimuli and testing algorithms. In conventional testing systems the subsystem complexity increases drastically as the DUT becomes more sophisticated, for instance, owing to the increased capacity of test data storage memory. The compact testing systems built according to the BIST architecture are used for the compressed representation of test information. Compact testing provides for test generation and response analysis according to the compact algorithms.

In engineering applications the algorithm complexity can be indirectly accessed from the complexity of applied hardware. For a compact algorithm, the implementation complexity is much lower than the complexity of test sequence carriers generated by the same algorithm.

7.2.1.1 Signature Analysis

Signature analysis was first developed in 1976, when Hewlett-Packard developed the signature analyzer and the analyzer for signal switching calculations. The highest diagnostic effect is achieved when the analyzers are applied to DUTs. The analysis of known technologies results in the following classification of DUTs:

1. Combinational circuit
2. Output-undependable finite state machine (FSM)
3. Input-undependable FSM
4. Input- and output-undependable FSM

Application of the signature analyzer permitted compression of long output sequences (output responses) into signature words. A concise engineering solution encouraged a number of theoretical and applied works dealing with the compression of sequences.

In early 1970s shift-register-based compact generators were extensively used as test stimuli sources. Application of compact generators reduced the hardware complexity at the cost of increased test sequence lengths. As no reliable diagnostic data concerning check completeness could be obtained, the tests generated were characterized as pseudo-random tests.

The open test systems are the most investigated class of compact systems; these are systems in which test generator (test pattern generator), DUT, and analyzer of output responses (response analyzer) are connected in series (Figure 7.1a). Further reduction of hardware complexity is achieved in closed systems where test generator, DUT, and analyzer of output responses create a closed loop (Figure 7.1b). Peculiar features of closed systems are caused by the effect of fault "multiplication" along the closed loop that enhances the detecting abilities. Since memory calls are absent during system operation, the testing process can be executed at the DUT working frequency. Application of closed systems for functional testing is discussed in Section 7.3 (System and Embedded Core Testing).

7.2.1.2 LFSR Models

In compact testing, test generators with linear feedback shift registers (LFSR) are the most universally employed. Such test generators, using very modest hardware, create sufficiently long sequences.

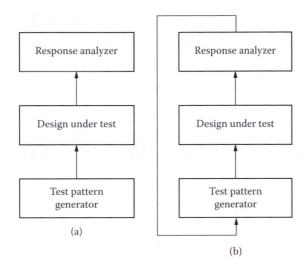

FIGURE 7.1
Open (a) and closed (b) testing systems.

We will first consider the algebraic elements designed for the synthesis of compact testing systems [3]. According to Gauss, a and b are *module p comparable* (p is an integer) if their difference divides by p without any residue. To express this, the following record is made:

$$a \equiv b \ (\text{mod } p).$$

The comparability relation is the equivalence relation, so the set of all integers is broken into the classes of mutually comparable numbers. The aggregate of such classes is a *ring*. If p is a prime number, the ring forms a *field*. The field that contains the finite number of q elements is called Galois field (GF(q)). The number of a finite field's elements is called its *order*.

Synthesis of the ring testing systems employs the notion of the polynomial of the z variable. The expression

$$g(z) = a_0 z^n + a_1 z^{n-1} + \cdots + a_n$$

is called the *polynomial over GF(p) field*, where a_i coefficients belong to GF(p). If $a_0 \neq 0$, it is said that g(z) has the degree of (deg g)n. The polynomial whose coefficient at the highest degree is equal to a unity is called the *monic* polynomial.

The totality of all polynomials over the GF(p) field forms a ring. For any g(z) and f(z) polynomials of this ring there exist such q(z) and r(z) polynomials that

$$g(z) = f(z) \ q(z) + r(z),$$

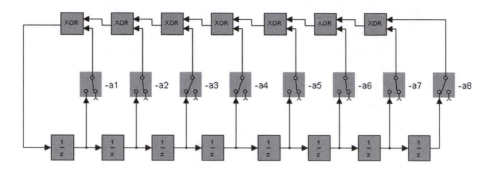

FIGURE 7.2
The structure of a single-adder test generator.

where the r(z) degree is less than the g(z) degree. Such a result is obtained when the polynomials are divided by the residual, similarly to the division of integers.

Now we examine test generators over the GF(p) finite field [3]. The elements of sequences generated by them take the values of the (0, 1,..., p-1) set. When p = 2, we have binary sequences for GF(2). Figure 7.2 and Figure 7.3 show the structures of single-adder and multi-adder test generators, respectively.

Test generators are realized in delay blocks, in gain blocks with a_1, ..., a_n coefficients, and in adders over mod p (for the p = 2, GF(2) case they become the XOR gates). The a_i coefficients belong to the (0, 1,..., p-1) set. When $a_i = 0$, the i-th connection between the delay element output and the adder input is absent in generator circuits. The single-adder and the multi-adder test generators start their performance following the initial setting of delay blocks (memory elements) to

$$X_n(0) = (x_1(0),..., x_n(0)); \ Y_n(0) = (y_1(0),..., y_n(0)) \text{ states.}$$

The $X_n(\tau)$, $Y_n(\tau)$ sets of outputs (states) of test generator delay blocks are used as test sets. In single-adder test generators test sets are generated by the matrix relation of

$$X_n(\tau + 1) = H \ X_n(\tau), \tag{7.1}$$

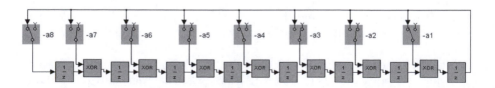

FIGURE 7.3
The structure of a multi-adder test generator.

where $X_n(\tau)$ is a column vector, and H is a square matrix:

$$H = \begin{Vmatrix} -a_1 & -a_2 & \cdots & -a_{n-1} & -a_n \\ 1 & 0 & \cdots & 0 & 0 \\ 0 & 1 & \cdots & 0 & 0 \\ \cdots\cdots\cdots\cdots\cdots\cdots\cdots\cdots\cdots \\ 0 & 0 & \cdots & 1 & 0 \end{Vmatrix}. \tag{7.2}$$

In multi-adder test generators test sets are generated by the

$$Y_n(\tau + 1) = H' \, Y_n(\tau) \tag{7.3}$$

matrix relation, where $Y_n(\tau)$ is a column vector,

$$H' = \begin{Vmatrix} -a_1 & 1 & 0 \cdots & 0 & 0 \\ -a_2 & 0 & 1 \cdots & 0 & 0 \\ \cdots\cdots\cdots\cdots\cdots\cdots\cdots \\ -a_{n-1} & 0 & 0 \cdots & 0 & 1 \\ -a_n & 0 & 0 \cdots & 0 & 0 \end{Vmatrix}. \tag{7.4}$$

The minus sign is omitted before the first row and first column elements of the Equation 7.2 and Equation 7.4 matrices, respectively, because the $a \equiv -a \pmod 2$ comparison is performed in the second-order field. It is the recursion relations that ensure the generation compactness in test sets. By proper choice of $a_1, \ldots, a_n$ coefficients, the test set sequences become strictly periodic. The T period is determined by the properties of the H, H' matrices — by the properties of their characteristic polynomial over the GF(p) field:

$$g(z) = a_0 z^n + a_1 z^{n-1} + \cdots + a_n \ (a_0 = 1).$$

In T units of time, a test generator recovers its initial state, having realized the cyclic path on the transition diagram. It is well-known that the zero initial state is a dead end and can generate only zero sets. Since H' is obtained by H transposition, then

$$g(z) = \det (H - zE) = \det (H' - zE)$$

defines one and the same T value for test generators:

$$X_n(T) = X_n(0); \ Y_n(T) = Y_n(0).$$

In this process the single-adder and multi-adder test generators are equivalent in their delay block outputs but nonequivalent in their states. As a consequence, similar initial states of test generators generate differing cyclic T-period sequences.

The T period is equal to the index responsible for g(z). Examining test generators with $a_1, \ldots, a_n$ coefficient sets that assign the irreducible g(z) polynomials over the GF(p) field, the binomial

$$z^e - 1 \; (e = 2^n - 1)$$

can be decomposed to the composition of irreducible polynomials

$$z^e - 1 = \Pi \, g_T(z), \; T \backslash e. \tag{7.5}$$

In Equation 7.5 the composition is performed by all e divisors of T. The g(z) polynomial, deg g = n, can be chosen out of the Equation 7.5 product. In practice, it is expedient to use the irreducible polynomial tables for n ≤ 34, as the complexity of obtaining Equation 7.5 grows dramatically with n.

A major function of the testing process is the analysis of tested objects' responses to test stimuli. The analysis aims at determining the serviceability or locating the fault place and type. The feedback analyzers serve exactly this purpose. In compact testing, the analyzers transform long output responses of a tested object into short sequences called signatures. Comparing the signatures obtained with reference signatures, one can make a decision on the fault-free DUT. Compact representation of responses results in memory savings in the testing systems. Similar to test generators, the analyzers are either counter or shift-register based.

We will further discuss the shift register–based feedback analyzers, using delay elements, amplifiers, and adders. Figure 7.4A–C and Figure 7.5A–E show the structures for single-adder and multi-adder analyzers, respectively. In contrast to the corresponding linear generator designs, the analyzers have data inputs, and the tested object responses arrive therein. If the signals are absent at these inputs, the analyzer acts as a generator.

The single-adder analyzer is used for the compression of a sequence that arrives at the adder input (Figure 7.4). If, using the feedback analyzer, a number of sequences should be compressed, they are pretransformed into a single sequence by summation or multiplexing. However, this may result either in the loss of diagnostic information or in the decrease of testing clock rate.

To study the analyzers' performance, their feedback structure is assigned by the coefficients of a

$$g(z) = a_0 z^n + a_1 z^{n-1} + \cdots + a_n \; (a_0 = a_n = 1) \tag{7.6}$$

polynomial over GF(2), and the analyzed binary sequence

$$\|b_i\| = b_k, \, b_{k-1}, \ldots, b_0$$

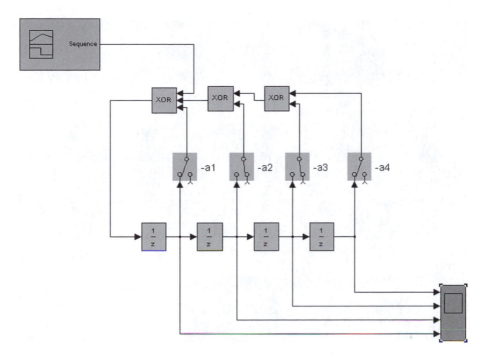

FIGURE 7.4A

The structure of a single-adder analyzer (n = 4), whose structure can be described by a g(z) = $z^4 + z^3 + 1$ polynomial.

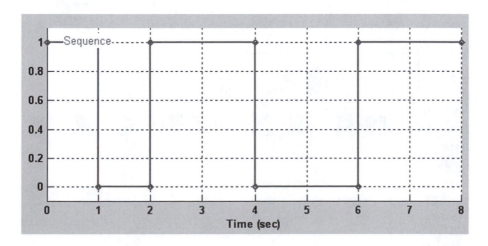

FIGURE 7.4B

The analyzed $\|b_i\| = 1, 0, 1, 1, 0, 0, 1, 1$ response.

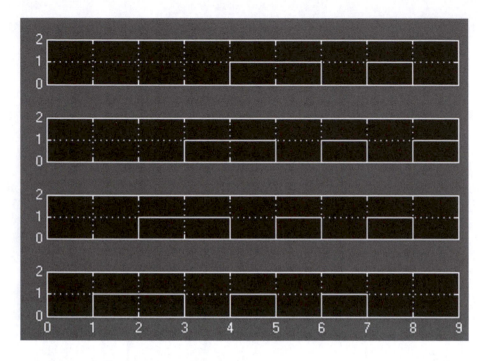

FIGURE 7.4C
Calculation result for a reference set (for a nondistorted input sequence).

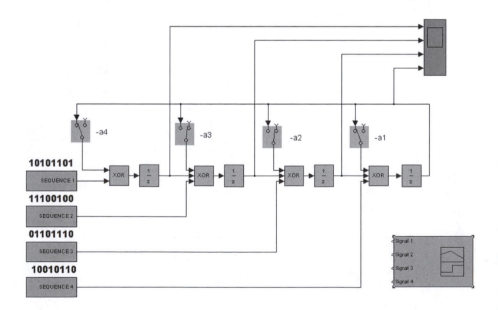

FIGURE 7.5A
The multi-adder analyzer (n = 4), whose structure can be described by a $g(z) = z^4 + z^3 + 1$ polynomial.

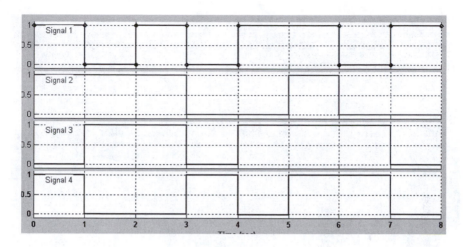

FIGURE 7.5B
The analyzed $\|b_{i,k}\|$ input sequences.

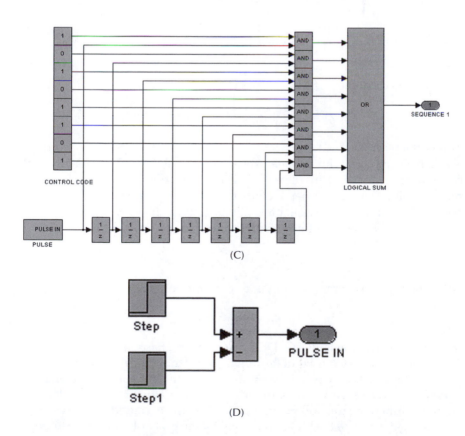

FIGURE 7.5 C,D
Block structure for input sequences of the Figure 7.5A analyzer.

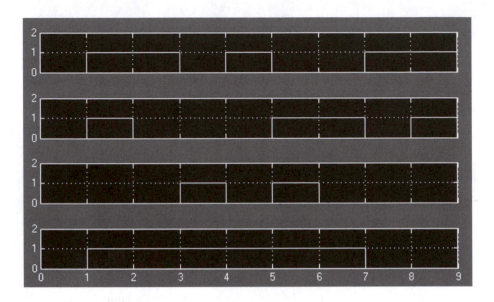

FIGURE 7.5E
Calculation result for a reference set E_n (for nondistorted input sequences).

is represented by the

$$h(z) = b_k z^k + b_{k-1}z^{k-1} + \cdots + b_0 \ (k > n)$$

polynomial.

It is believed that with the analyzer's zero initial state, the elements of the $\|b_i\|$ sequence arrive at the input in decreasing order of indices. Following register shift, each testing step records in its first digit the value that depends on the input state, whereas all other digits assume the previous values.

Analyzer performance finalizes in the register-generated set of

$$D_n = (d_1, \ldots, d_n).$$

After the D_n set is compared to the reference one,

$$E_n = (e_1, \ldots, e_n),$$

the conclusion on the fault-free DUT can be made, based on its $\|b_i\|$ response. The reference E_n set can be determined by mathematical simulation of the analyzer function or by physical simulation of a fault-free DUT and analyzer.

When the reference E_n set is found from the mathematical simulation of analyzer function, the following operations are executed: (a) division of decreasing degrees of the $h(z)$ polynomial by the $g(z)$ polynomial, resulting in

$$h(z) = g(z) q(z) + r(z),$$

where q(z) is a quotient, and the residual looks like

$$r(z) = c_n z^{n-1} + c_{n-1} z^{n-2} + \cdots + c_1.$$

(b) the

$$E_n = \| e_1 \cdots e_n \|$$

matrix is calculated by linear transformation of the

$$C_n = \| c_1 \cdots c_n \|$$

matrix, incorporating the residue coefficients,

$$E_n = C_n A^{-1}, \tag{7.7}$$

where A^{-1} is an inverse matrix with regard to the nondegenerate one:

$$A = \begin{Vmatrix} a_{n-1} & a_{n-2} & \cdots & a_1 & a_0 \\ a_{n-2} & a_{n-3} & \cdots & a_0 & 0 \\ & & \cdots & & \\ a_0 & 0 & \cdots & 0 & 0 \end{Vmatrix}.$$

Calculation of E_n can also be made from the functions of analyzer transitions and outputs.

Example 1 [3]

Let a polynomial that describes the analyzer structure (Figure 7.4A)

$$g(z) = z^4 + z^3 + 1$$

and the analyzed sequence (Figure 7.4b)

$$\| b_i \| = 1, 0, 1, 1, 0, 0, 1, 1$$

be known. The E_n matrix should be found. Representing $\| b_i \|$ as the

$$h(z) = z^7 + z^5 + z^4 + z + 1$$

polynomial, we have from the polynomial division algorithm

$$h(z) = g(z)(z^3 + z^2 + 1) + (z^2 + 1),$$

whence it follows that

$$\mathbf{C}_n = \| \, 0 \ 1 \ 1 \ 0 \, \|.$$

Having determined the

$$\mathbf{A} = \begin{Vmatrix} 0 \ 0 \ 1 \ 1 \\ 0 \ 1 \ 1 \ 0 \\ 1 \ 1 \ 0 \ 0 \\ 1 \ 0 \ 0 \ 0 \end{Vmatrix}, \quad \mathbf{A}^{-1} = \begin{Vmatrix} 0 \ 0 \ 0 \ 1 \\ 0 \ 0 \ 1 \ 1 \\ 0 \ 1 \ 1 \ 1 \\ 1 \ 1 \ 1 \ 1 \end{Vmatrix}$$

matrices from Equation 7.6, we have (Figure 7.4C)

$$\mathbf{E}_n = \| 0 \ 1 \ 1 \ 0 \| \begin{Vmatrix} 0 \ 0 \ 0 \ 1 \\ 0 \ 0 \ 1 \ 1 \\ 0 \ 1 \ 1 \ 1 \\ 1 \ 1 \ 1 \ 1 \end{Vmatrix} = \| 0 \ 1 \ 0 \ 0 \|.$$

The multi-adder n-digit analyzer (Figure 7.5A) is intended for parallel analysis of no more than n output responses of a DUT that arrive at the adders' inputs. In contrast to a single-adder analyzer, a multi-adder analyzer does not need any preliminary transformation of sequences.

As before, the analysis is made at the analyzer's zero starting state. Each testing step records the results of summation of input values and output values of corresponding register digits in the analyzer's memory elements.

We can describe the input stimulus, Ω, as a set of ordered elements representing analyzed responses of a DUT:

$$\Omega = \{\|b_{i,1}\|, \|b_{i,2}\|, \ldots, \|b_{i,n}\|\},$$

where

$$\|b_{i,j}\| = b_{k,j}, \ b_{k-1,j}, \ldots, \ b_{0,j} \ (j = 1, \ldots, n; \ k > n).$$

The elements of each $\|b_{i,j}\|$ sequence enter the input in decreasing order of their indices. Following Ω application, the analyzer state has the value of $\mathbf{S}_n = (s_1, \ldots, s_n)$. We can represent the set as the sum of

$$\Omega = \omega_1 \oplus \omega_2 \oplus \cdots \oplus \omega_n,$$

where

$$\omega_1 = \{\|b_{i,j}\|, 0, \ldots, 0\};$$

$$\omega_2 = \{0, \|b_{i,j}\|, \ldots, 0\};$$

$$\ldots$$

$$\omega_n = \{0, 0, \ldots, \|b_{i,j}\|\}.$$

Therefore, the stimulus is simultaneous for all inputs breaks into the totality of time-consecutive stimuli for each j-th input. We will denote the result of the ω_i application recorded at the zero-initial-state register as the $V_{n,j} = (v_{1,j}, \ldots, v_{n,j})$ set. Then by virtue of the fact that the analyzer is a linear machine and thus possesses the superposition property, we have

$$s_i = v_{i,1} \oplus v_{i,2} \oplus \cdots \oplus v_{i,n} \ (i = 1, \ldots, n). \tag{7.8}$$

The Equation 7.8 equality is needed to calculate the $E_n = (e_1, \ldots, e_n)$ reference set, necessary for comparison with the S_n actual set.

Assuming that the analyzer feedback is characterized by the g(z) polynomial of the Equation 7.6 kind and the polynomial representation of the $\|b_{i,j}\|$ sequence looks like

$$h_j(z) = b_{k,j} z^n + b_{k-1,j} z^{n-1} + \cdots + b_{0,j}, \tag{7.9}$$

we can find $V_{n,j}$ sets for a fault-free object. As long as by ω_j application to the j-th input the analyzer multiplies the $h_j(z)$ polynomial by z^{j-1} and then divides the product obtained by g(z), we have

$$h_1(z) = g(z) \, q_1(z) + r_1(z);$$

$$zh_2(z) = g(z) \, q_2(z) + r_2(z); \tag{7.10}$$

$$\ldots$$

$$z^{n-1} h_n(z) = g(z) \, q_n(z) + r_n(z).$$

Unlike similar calculations for a single-digit analyzer, no linear transformation of Equation 7.7 is necessary here. After the Equation 7.10 system is generated, the residuals of

$$r_1(z) = v_{n,1} z^{n-1} + v_{n-1,1} z^{n-2} + \cdots + v_{1,1};$$

$$r_2(z) = v_{n,2} z^{n-1} + v_{n-1,2} z^{n-2} + \cdots + v_{1,2};$$

$$\ldots$$

$$r_n(z) = v_{n,n} z^{n-1} + v_{n-1,n} z^{n-2} + \cdots + v_{1,n};$$

assist in defining the E_n reference set components:

$$e_1 = v_{1,1} \oplus v_{1,2} \oplus \cdots \oplus v_{1,n};$$

$$\cdots \tag{7.11}$$

$$e_{n-1} = v_{n-1,1} \oplus v_{n-1,2} \oplus \cdots \oplus v_{n-1,n};$$

$$e_n = v_{n,1} \oplus v_{n,2} \oplus \cdots \oplus v_{n,n}.$$

Thus, the $E_n = (e_1, \ldots, e_n)$ response to the Ω stimulus is calculated from the reference responses to ω_j stimuli.

Example 2 [3]
Let there be a

$$g(z) = z^4 + z^3 + 1$$

polynomial that describes the Figure 7.5A analyzer structure and the analyzed sequences (Figure 7.5B)

$$\|b_{i,1}\| = 1, 0, 1, 0, 1, 1,0, 1;$$

$$\|b_{i,2}\| = 1, 1, 1, 0, 0, 1, 0, 0;$$

$$\|b_{i,3}\| = 0, 1, 1, 0, 1, 1, 1, 0;$$

$$\|b_{i,4}\| = 1, 0, 0, 1, 0, 1, 1, 0.$$

The E_n reference set should be found.

If we represent the $\|b_{i,j}\|$ sequences as the polynomials of Equation 7.9 and perform the multiplication and division operations over them, according to Equation 7.11, we have (Figure 7.5E):

$$h_1(z) = g(z) (z^3 + z^2) + 1;$$

$$z\, h_2(z) = g(z) (z^4 + z^2 + z) + (z^3 + z^2 + z);$$

$$z^2\, h_3(z) = g(z) (z^4 + z + 1) + (z + 1);$$

$$z^3\, h_4(z) = g(z) (z^6 + z^5 + z^4 + z^2 + z + 1) + (z^3 + z^2 + z + 1).$$

Then, from Equation 7.11, we find $E_n = (1100)$ (Figure 7.5E).

It can be seen from Figure 7.5A that in this case the input sequence sources are the sequence units rather than the Signal Builder Block, and their internal structures are given in Figure 7.5C,D. This is caused by the instability of the Signal Builder Block operation in such circuits.

As the section summary, we give the example of an elementary open compact-testing system devised in compliance with the BIST architecture (Figure 7.1A, Figure 7.6A). The system contains two LFSRs: LFSR1, serving as a test generator (Figure 7.6B), and LFSR2, serving as response analyzer (Figure 7.6C). The circuit for a fault-free DUT is shown in Figure 7.6D. A time graph of the test generator operation is shown in Figure 7.6E. The time graph of the response analyzer operation with a DUT reference response is shown in Figure 7.6F. It can be seen that the three-bit code 110 is the reference signature (Figure 7.6A,F). When DUT contains a fault (Figure 7.6G), the response analyzer calculates the 000 signature (Figure 7.6H).

7.2.2 Scan Cell Operations

The hardware for compact testing systems comprises universal facilities that — depending on the need — can serve as test generators or as output response analyzers. As the generator and the analyzer circuit engineering is common, their joint compact realization as a universal observer is also possible. Figure 7.7 and Figure 7.8 demonstrate the structures for universal single-adder and multi-adder controllers. With CONTROL = 0 signal, the observers operate like generators; with CONTROL = 1, like analyzers.

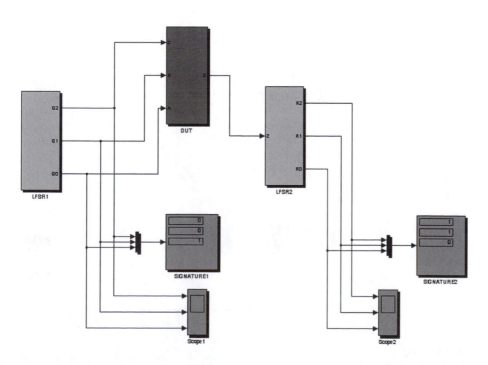

FIGURE 7.6A
Elementary open BIST architecture–based compact-test system.

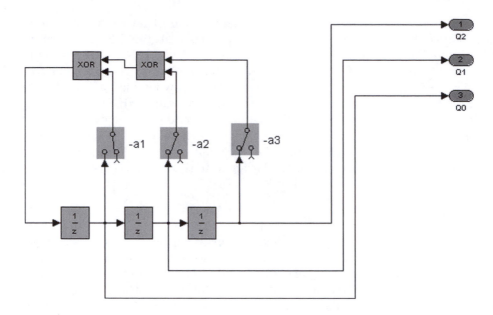

FIGURE 7.6B
The LFSR1 structure, a single-adder test generator (n = 3).

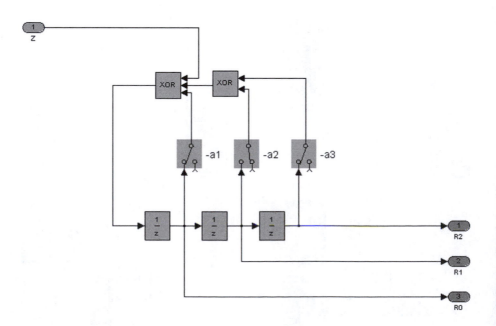

FIGURE 7.6C
The LFSR2 structure, a single-adder response analyzer (n = 3).

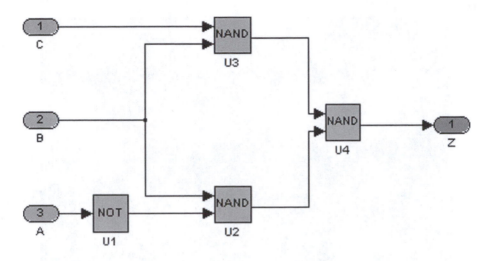

FIGURE 7.6D
Structure of a fault-free DUT.

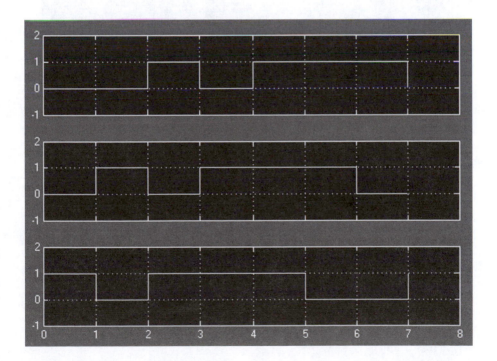

FIGURE 7.6E
Time diagram of test generator operation.

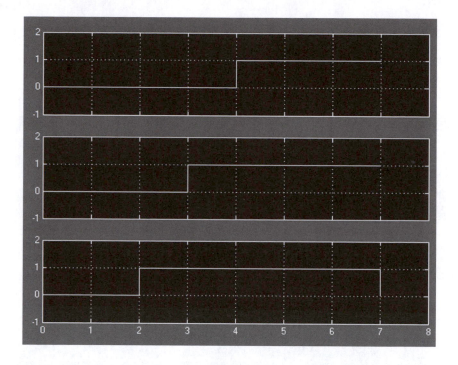

FIGURE 7.6F
Time diagram of the response analyzer operation for a fault-free DUT.

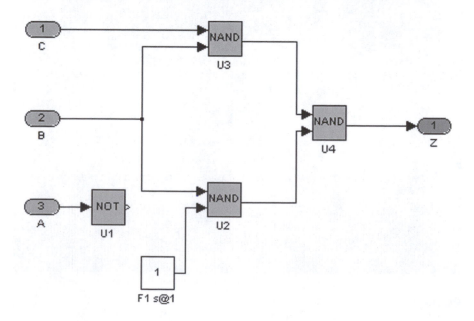

FIGURE 7.6G
The structure of an F1 fault–containing DUT.

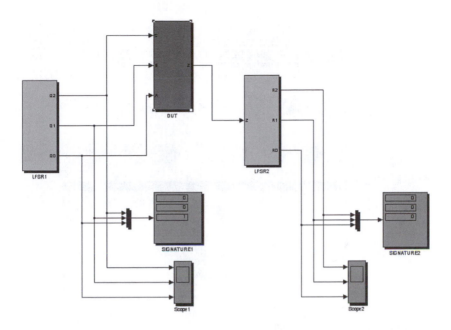

FIGURE 7.6H
Calculation of the 000 signature, corresponding to the Figure 7.6G F1 fault.

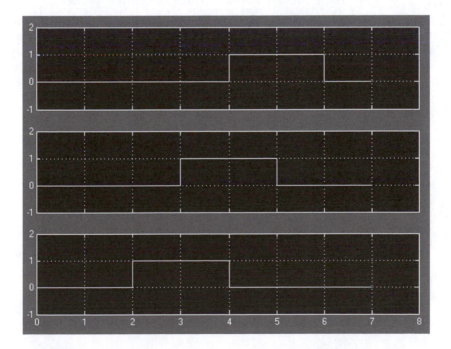

FIGURE 7.6I
Time diagram of the response analyzer operation for the F1 fault–containing DUT.

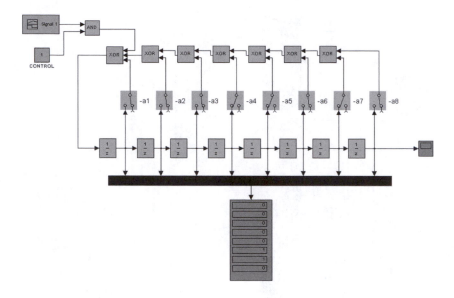

FIGURE 7.7
A single-adder multipurpose controller.

7.2.2.1 Built-In Logic Block Observer (BILBO) Register Model

The design of BILBO is the most appropriate in signature analysis [1,4] (Figure 7.9). By simple adjustment of its scanning cells, the controller permits performance of either response analysis (i.e., serving as an output response analyzer)

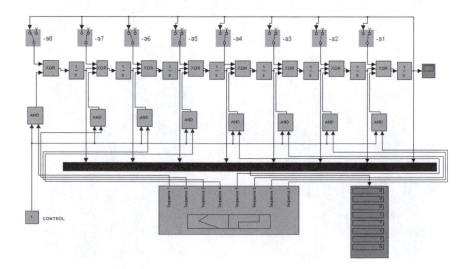

FIGURE 7.8
A multi-adder multipurpose controller.

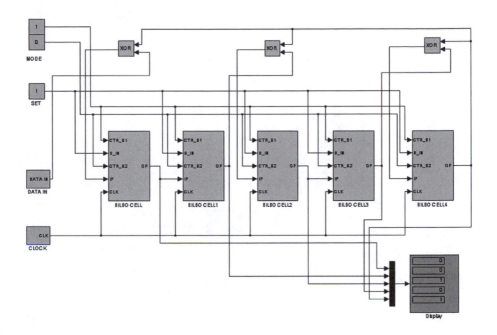

FIGURE 7.9A
A five-bit BILBO register model.

using signature analysis or test generation (serving as a test pattern generator). The controller's low hardware complexity stems from the ability of LFSR to act in both functions. The base BILBO register can function in four operation modes set by the CTR_B1, CTR_B2 control signals (Figure 7.9, Table 7.1).

Using the BILBO register, we can view the results of operation performance by DUT via XOR gates incorporated into LFSR and connected to a set of nodes in the device. Input data (the DUT responses) and the LFSR current content are the operands of XOR gates that perform the operand convolutions. At the end of its test sequence, the BILBO register contains a signature (syndrome) for the sequence that can be compared to the reference signature. The BILBO register can not only represent the random pattern generator or signature analyzer but be employed as a normal register and scan register, depending

TABLE 7.1

BILBO Register Modes

CTR_B1	CTR_B2	Mode
0	0	Linear shift (scan)
1	0	Signature analysis
1	1	Data (complemented) latch (normal)
0	1	Reset

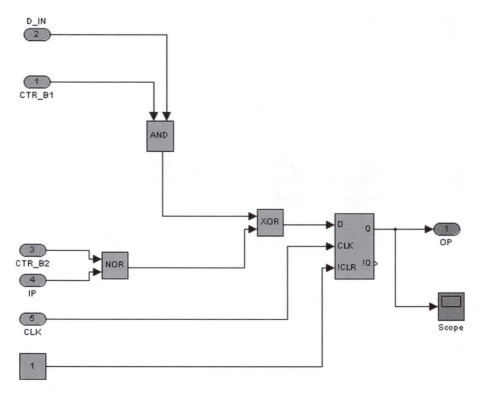

FIGURE 7.9B
BILBO register cell (digit) structure.

on the control signal values. Using the scanning option, the test sequence can be recorded in the input BILBO register 1 while initializing the output BILBO Register 2. Then BILBO register 1 and BILBO register 2 will operate in pattern generation and signature analysis modes, respectively. When testing is over, the signature is read from BILBO Register 2, using scan mode.

7.3 System and Embedded Core Testing

7.3.1 Functional Testing

Ring testing systems display consequent development of closed systems [3]. In ring testing systems, the generator and analyzer functions are time- and space-coincident, the system topology resembles a ring, and similar system models are described by the polynomial ring algebras and by the ring (cyclic) graphs, which generated the term *ring testing*.

During testing, the fault-free DUT passes through its states in a cyclic path. That is why the conclusion of the fault-free DUT is made from the comparison of its initial and final states. Redundant hardware of ring testing systems directly depends on the DUT linearity and nonlinearity properties. Since the generator and analyzer functions are combined, the redundancy of test systems becomes insignificant for a large variety of DUTs. Application of linear ring test systems is most expedient for the case of embedded functional testing. Such systems look like strictly periodical systems (without preperiods) with a cyclic operation nature.

Let DUT be a combinational discrete device. In the ring testing system for combinational devices, the functions of test generator and DUT output response analyzer in time and in space can be integrated in the easiest way.

Let a combinational device have n inputs and m outputs and be described by the system of Boolean functions:

$$y_i = f_i(x_1, \ldots, x_n) \ (i = 1, \ldots, m).$$

In this case, the ring testing linear system (Figure 7.10A–E) contains:

1. A combinational CORRECTION UNIT
2. An n-bit shift register (pseudo-random binary sequence generator [PRBSG])
3. Three XOR gates

The correction unit is intended for the generation of linear feedback function

$$\varphi(x_1, \ldots, x_n) = a_1 \, x_1 \oplus \cdots \oplus a_n \, x_n. \tag{7.12}$$

The feedback from the XOR gate output to the PRBSG input combines the register functions as the test generator and an in-series analyzer of the DUT output responses to the tests.

The combinational device testing is performed at $\tau = 1, 2, \ldots$ moments and is described by the recurrent ratio of

$$\varphi(\tau) = \sum_{i=1}^{n} a_i \, \varphi(\tau - i). \tag{7.13}$$

Here and below, the summation is made from module 2. The initial conditions for Equation 7.13 are given by the set of values

$$\varphi(-1) = x_1 \, (0), \ \varphi(-2) = x_2(0), \ \varphi(-3) = x_3 \, (0), \ldots, \ \varphi(-n) = x_n \, (0),$$

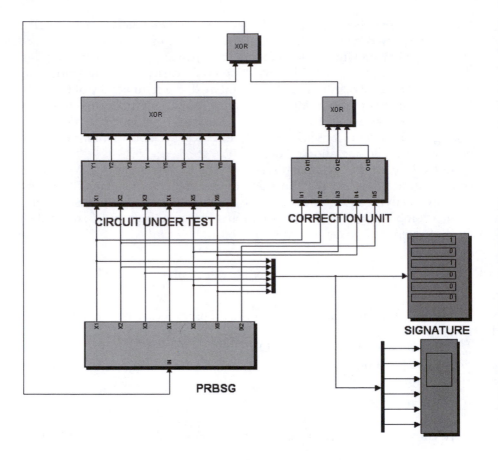

FIGURE 7.10A
Linear ring testing system structure.

corresponding to the PRBSG register initial state. When a strictly periodic ring testing system exists for the $\{\varphi(\tau)\}$ sequence (the solution of Equation 7.13), there is such integer value of T that $\varphi(\tau) = \varphi(\tau + T)$.

To analyze the periodicity of the ring testing system, the properties of a polynomial ring over GF(2) are used. The period of the ring testing system is known to be identical to T, to which the irreducible feedback polynomial g(z) belongs:

$$g(z) = \sum_{i=1}^{n} a_i \, z^{n-i} \ (a_0 = a_n = 1). \tag{7.14}$$

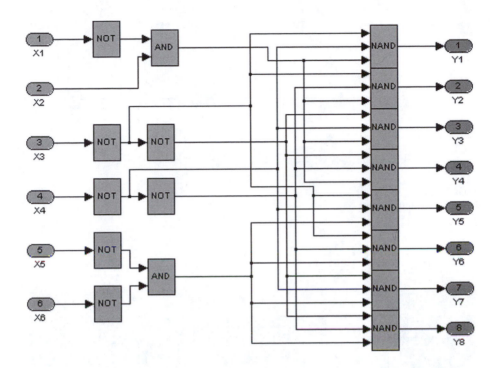

FIGURE 7.10B
The structure of the DUT combinational device.

The synthesis of the T-periodic ring testing system is based on the correction unit synthesis. The feedback function of Equation 7.12), defined by the set of a_i coefficients of the Equation 7.14 polynomial assigns the system synthesis.

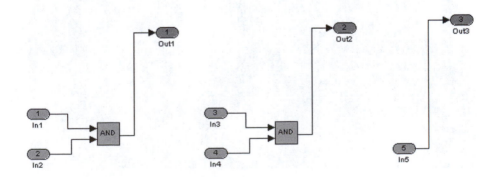

FIGURE 7.10C
Realization of a correction unit.

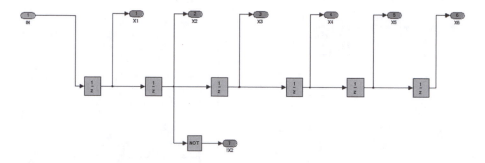

FIGURE 7.10D
The PRBSG shift register.

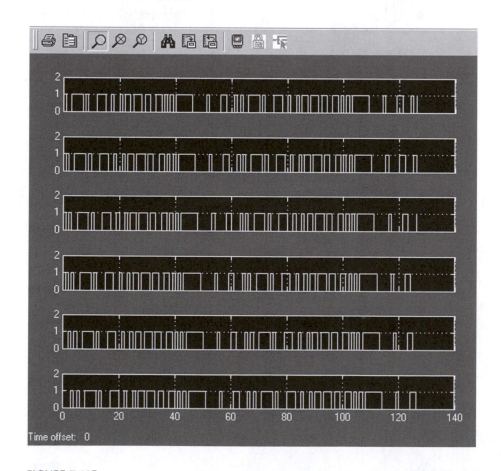

FIGURE 7.10E
Time state diagrams of a six-digit PRBSG shift register.

To construct the correction unit, the DUT sum of functions is found:

$$\Phi(x_1,\ldots,x_n) = \sum_{i=1}^{n} f_i(x_1,\ldots,x_n).$$

The sum is then expressed via the Zhegalkin polynomial [3,5]:

$$\Phi(x_1,\ldots,x_n) = \Sigma\, b_{j1\ldots js}\, x_{j1}\cdots x_{js},$$

where $b_{j1\ldots js} = 0$ or 1, and the summation is made for all $(j1,\ldots,js)$ subsets of the $(1,\ldots,n)$ set.

In Zhegalkin algebra the following connectives are used:

1. Logical multiplication (conjunction)
2. Addition by module 2 (exclusive OR, XOR)
3. Constant 1 ("the truth")

The set of operations is complete, that is, each logical algebra function can be represented as a superposition of said operations. Moreover, in Zhegalkin algebra, each logical algebra $f_i(x_1,\ldots,x_n)$ function is unambiguously represented as a polynomial wherein each x_i variable can be no higher than the first power, and the elements of the GF(2) field are its coefficients. The possibility of such a representation by the reduced polynomials stems from the existence of the Lagrange interpolation formula, which in this case is very simple:

$$f(x_1,\ldots,x_n) = \sum_{(a_1,\ldots,a_n)} f(a_1,\ldots a_n)(x_1 + a_1 + 1)\cdots(x_n + a_n + 1).$$

The Boolean connectives, disjunction, and negations in Zhegalkin algebra are expressed as

$$x_1 \vee x_2 = x_1 \oplus x_2 \oplus x_1\, x_2, \tag{7.15}$$

$$\overline{x} = x \oplus 1.$$

The Zhegalkin algebra is also called the Boolean ring. The operations on reduced polynomials are performed just like on the common polynomials with integer coefficients. Then in the result obtained, all x_i^m variables with $m > 0$ are replaced by x_i, and the mononomial coefficients are replaced by their residuals by module 2. The affinity of Zhegalkin algebra and conventional elementary algebra of polynomials can explain its preference from the methodological viewpoint.

The representation of the $\Phi(x_1,\ldots, x_n)$ function as a Zhegalkin polynomial reveals the properties associated with DUT linearity and nonlinearity (in this case the combinational device). The assignment for the synthesis of a correction unit with a XOR output gate is given as follows:

$$F(x_1,\ldots, x_n) = \varphi(x_1,\ldots, x_n) \oplus \Phi(x_1,\ldots, x_n).$$

Since the $F(x_1,\ldots, x_n)$ function is expressed as the Zhegalkin polynomial, then in a general case, for the realization of its constituents, the correction unit includes direct connections and AND gates, whose outputs are connected to the XOR gate inputs.

During operation of the ring testing system, binary test sets X_n are fed to the DUT inputs according to the recurrent ratio of

$$X_n(\tau + 1) = H \, X_n(\tau) = H^{\tau+1} \, X_n(0), \, (\tau = 0, 1,\ldots); \qquad (7.16)$$

where $X_n(\tau) = (x_1(\tau),\ldots, x_n(\tau))$ is column vector, and

$$H = \begin{Vmatrix} a_1 & a_2 & \cdots & a_{n-1} & a_n \\ 1 & 0 & \cdots & 0 & 0 \\ 0 & 1 & \cdots & 0 & 0 \\ \cdots & \cdots & \cdots & \cdots & \cdots \\ 0 & 0 & \cdots & 1 & 0 \end{Vmatrix}$$

is accompanying matrix of an irreducible polynomial $g(z)$.

When the DUT — combinational device — and the ring testing system itself are fault-free, we have

$$X_n(T) = X_n(0),$$

which follows from Equation 7.16, taking into account that $H^T = E$. This makes it possible to ascertain the absence of faults in DUT by observing similar states of the PRBSG register outputs before and after testing.

Example 3 [3]

Let DUT be a combinational device whose structure is given in Figure 7.10B. The DUT has n = 6 inputs and m = 8 outputs. The ring test system with T = 63 should be constructed.

First, the DUT output functions should be expressed via Zhegalkin polynomials. For instance, the $Y1 = \overline{(\overline{x_1 x_2}) x_3 x_4}$ output function is expressed by the Zhegalkin polynomial in the course of the following calculations (using the Equation 7.15 ratios).

$$Y1 = \overline{(\overline{x_1 x_2}) x_3 x_4} = \overline{((1+x_1)x_2)} \vee \overline{(1+x_3)} \vee \overline{(1+x_4)} = (x_2 + x_1 x_2) \vee (x_3) \vee (x_4)$$
$$= (1 + x_2 + x_1 x_2) \vee x_3 \vee x_4 = (1 + x_2 + x_1 x_2) \vee (x_3 + x_4 + x_3 x_4)$$
$$= 1 + x_2 + x_1 x_2 + x_3 + x_4 + x_3 x_4 + x_3 + x_4 + x_3 x_4 + x_2 x_3 + x_2 x_4 + x_2 x_3 x_4 + x_1 x_2 x_3$$
$$+ x_1 x_2 x_4 + x_1 x_2 x_3 x_4 = 1 + x_2 + x_1 x_2 + x_2 x_3 + x_2 x_4 + x_2 x_3 x_4 + x_1 x_2 x_3 + x_1 x_2 x_4 + x_1 x_2 x_3 x_4.$$

In the same manner, the Zhegalkin polynomials assist in expressing the remaining functions of the Y2... Y8 outputs:

$$Y2 = 1 + x_2 x_4 + x_2 x_3 x_4 + x_1 x_2 x_4 + x_1 x_2 x_3 x_4;$$

$$Y3 = 1 + x_2 x_3 + x_2 x_3 x_4 + x_1 x_2 x_3 + x_1 x_2 x_3 x_4;$$

$$Y4 = 1 + x_2 x_3 x_4 + x_1 x_2 x_3 x_4;$$

$$Y5 = x_3 + x_4 + x_5 + x_6 + x_3 x_5 + x_4 x_5 + x_5 x_6 + x_3 x_6 + x_4 x_6 + x_3 x_4$$
$$+ x_3 x_4 x_5 + x_4 x_5 x_6 + x_3 x_5 x_6 + x_3 x_4 x_6 + x_3 x_4 x_5 x_6;$$

$$Y6 = 1 + x_4 + x_3 x_4 + x_4 x_5 + x_4 x_6 + x_4 x_5 x_6 + x_3 x_4 x_6 + x_3 x_4 x_5 + x_3 x_4 x_5 x_6;$$

$$Y7 = 1 + x_3 + x_3 x_4 + x_3 x_5 + x_3 x_6 + x_3 x_4 x_6 + x_3 x_5 x_6 + x_3 x_4 x_5 + x_3 x_4 x_5 x_6;$$

$$Y8 = 1 + x_3 x_4 + x_3 x_4 x_5 + x_3 x_4 x_6 + x_3 x_4 x_5 x_6.$$

Now we can sum the functions up:

$$\Phi = \sum_{I=1}^{8} YI = 1 + x_2 + x_5 + x_6 + x_1 x_2 + x_5 x_6.$$

On the other hand, when the $z^{63} - 1$ binomial is decomposed from Equation 7.5, we have the following GF(2) irreducible polynomials of the sixth power:

$$g_{63}(z) = z^6 + z + 1;$$

$$g_{63}(z) = z^6 + z^4 + z^3 + z + 1;$$

$$g_{63}(z) = z^6 + z^5 + 1;$$

$$g_{63}(z) = z^6 + z^5 + z^2 + z + 1;$$

$$g_{63}(z) = z^6 + z^5 + z^3 + z^2 + 1;$$

$$g_{63}(z) = z^6 + z^5 + z^4 + z + 1;$$

$$g_{21}(z) = z^6 + z^4 + z^2 + z + 1;$$

$$g_{21}(z) = z^6 + z^5 + z^4 + z^2 + 1;$$

$$g_9(z) = z^6 + z^3 + 1.$$

Let us assume that the $g_{63}(z) = z^6 + z^5 + 1$ polynomial belonging to $T = 63$ is chosen. Then the feedback function of the

$$\varphi = x_5 \oplus x_6$$

type should be taken, and

$$F = \varphi \oplus \Phi = 1 \oplus x_2 \oplus x_1 x_2 \oplus x_5 x_6 = \bar{x}_2 \oplus x_1 x_2 \oplus x_5 x_6$$

will serve as the correction unit assignment. Therefore, to realize the correction unit (Figure 7.10C), we need two AND gates and one direct connection to the $\bar{x}_2$ output of the PRBSG shift register (Figure 7.10D).

Now DUT testing in the class of single constant faults can be performed at any initial state of the PRBSG register, except the zero state.

If in 63 operation tacts, no DUT faults are found, our ring testing system (its shift register PRBSG) resumes its initial state (Figure 7.10E). Hence it follows that a fault-free ring testing system and a shift-register-based single-adder generator are equivalent in their XOR gate outputs, because the similar feedback functions are realized there. In the process, the given ring testing system has a drawback: all DUT faults at the even XOR gate inputs are not necessarily displayed at the gate output.

We will now assume that DUT is an output-independent memory machine, that is, a combinational circuit with time delays transformed into testable shape via the break of its feedbacks. Such DUT is described by the following expression:

$$YI = f_i(x_1(\tau),\ldots, x_n(\tau), x_1(\tau - 1),\ldots, x_n(\tau - 1),\ldots, x_1(\tau - \mu),\ldots, x_n(\tau - \mu))$$

$$(I = 1,\ldots, m; \tau = 0,1,\ldots), \tag{7.17}$$

where n is the number of DUT inputs, m is the number of DUT outputs, and μ is the memory depth by x.

The ring testing system for such DUT contains (Figure 7.11A):

1. An output-independent CORRECTION UNIT
2. An n-bit shift register PRBSG
3. A μ-bit auxiliary shift register PRBSG1
4. The XOR gates

The system feedback function looks like

$$\varphi(x_1,\ldots, x_r) = a_1 x_1 \oplus \cdots \oplus a_n x_n \oplus a_{n+1} x_{n+1} \cdots \oplus a_r x_r, \ (r = n + \mu), \tag{7.18}$$

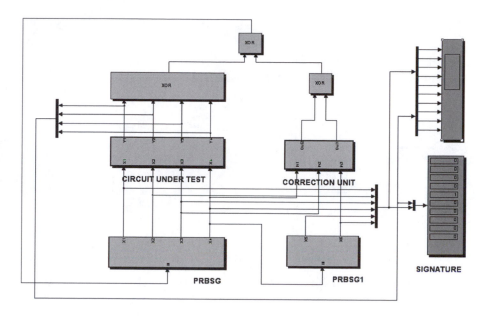

FIGURE 7.11A

Ring testing system for the DUT represented by an output-independent FSM.

where $x_1, \ldots, x_n$ are the variables of the DUT inputs and the correction unit that arrive from the PRBSG register outputs, and $x_{n+1}, \ldots, x_r$ are additional variables of the correction unit inputs that arrive from the PRBSG1 register outputs (Figure 7.11A).

The system operation is described by

$$\varphi(\tau) = \sum_{i=1}^{r} a_i \, \varphi(\tau - i),$$

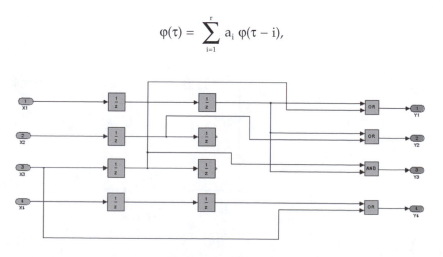

FIGURE 7.11B

DUT represented by an output-independent FSM.

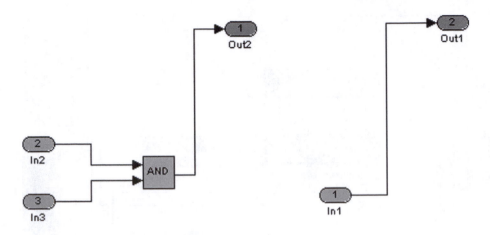

FIGURE 7.11C
Correction unit structure.

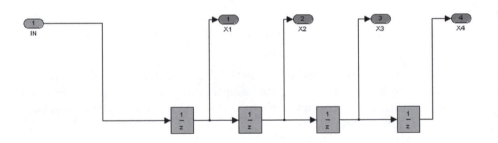

FIGURE 7.11D
PRBSG shift register structure.

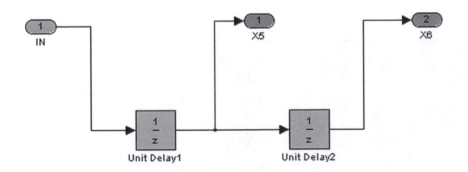

FIGURE 7.11E
PRBSG1 auxiliary shift register structure.

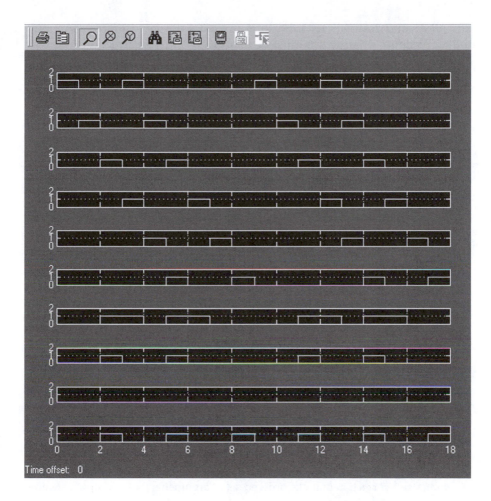

FIGURE 7.11F
Test results (the six initial signals are the PRBSG and PRBSG1 outputs).

and its period is equal to the exponent of the polynomial

$$g(z) = \sum_{i=1}^{r} a_i z^{r-i} \ (a_0 = a_r = 1). \tag{7.19}$$

of the polynomial ring that GF(2) belongs to.

To synthesize the ring testing system of the T period, the feedback function of the Equation 7.18 type is given. As far as the PRBSG and the XOR elements have the same destination as the combinational device testing system, only the correction unit structure must be examined. Such a correction unit with output XOR gate must realize the function that can be represented as a sum of linear and nonlinear summands [3]. For their generation, the memory that is constructed of the PRBSG1 register delay elements, the combinational unit that comprises the XOR gate for a linear summand, and the AND gate for a nonlinear summand, are required.

Example 4 [3]

Let DUT have $n = 4$ inputs and $m = 4$ outputs and the number of its delay cascades be $\mu = 2$. The DUT is described by the following ratios (Figure 7.11B):

$$Y1(\tau) = x_1(\tau - 2) \vee x_3(\tau - 1);$$

$$Y2(\tau) = x_1(\tau - 2) \vee x_2(\tau - 1);$$

$$Y3(\tau) = x_1(\tau - 2) \vee x_3(\tau - 1);$$

$$Y4(\tau) = x_3(\tau) \vee x_4(\tau - 2).$$

The ring testing system of $T = 9$ period should be constructed. In this case, the feedback polynomial is given as

$$g_9(z) = z^6 + z^3 + 1.$$

The following function corresponds to this polynomial:

$$\varphi(\tau) = \varphi(\tau - 3) \oplus \varphi(\tau - 6).$$

To synthesize a correction unit, the ordered set of arguments for the output functions is first enumerated:

$$YI(\tau) = f_i(x_1(\tau), x_2(\tau), x_3(\tau), x_4(\tau),$$

$$x_1(\tau - 1), x_2(\tau - 1), x_3(\tau - 1), x_4(\tau - 1),$$

$$x_1(\tau - 2), x_2(\tau - 2), x_3(\tau - 2), x_4(\tau - 2)) \Rightarrow f_i(x_1, x_2, \ldots, x_{12}).$$

Expressing the output functions by the Zhegalkin polynomials and summing them up, we have (passing from $x_m[\tau - k]$ to $\varphi[\tau - (m + k)]$)

$$\Phi = x_3 \oplus x_6 \oplus x_7 \oplus x_{12} \oplus x_6 x_9 \oplus x_3 x_{12}$$

or

$$\Phi(\tau) = \varphi(\tau - 3) \oplus \varphi(\tau - 3) \oplus \varphi(\tau - 4) \oplus \varphi(\tau - 6) \oplus \varphi(\tau - 3)\,\varphi(\tau - 3) \oplus \varphi(\tau - 3)\,\varphi(\tau - 6)$$

$$= \varphi(\tau - 3) \oplus \varphi(\tau - 4) \oplus \varphi(\tau - 6) \oplus \varphi(\tau - 3)\,\varphi(\tau - 6).$$

Thus, the function of

$$F(\tau) = \varphi(\tau) \oplus \Phi(\tau) = \varphi(\tau - 4) \oplus \varphi(\tau - 3)\,\varphi(\tau - 6)$$

is the assignment for the correction unit synthesis.

Hence it follows that the correction unit comprises (Figure 7.11C) a direct connection of the fourth digit output in the PRBSG shift register and the XOR gate input (Figure 7.11A), as well as the AND gate whose inputs are connected to the third digit outputs of PRBSG shift register and to the sixth digit of the auxiliary PRBSG1 shift register (Figure 7.11A). The structures of the PRBSG shift register and of the auxiliary PRBSG1 shift register are given in Figure 7.11D and Figure 7.11E, respectively.

If in nine operation pitches, no DUT faults are found, our ring testing system (its shift registers are PRBSG and PRBSG1) resumes its initial state (Figure 7.11F).

Now we can examine the construction of such a ring testing system for a DUT, where DUT is an input-independent FSM; it does not have any data inputs at which the test stimuli could have arrived. If a DUT is not redundant hardware, any of its faults are displayed at the DUT's outputs during its performance. In this manner, the performance mode is at the same time the testing mode for any autonomous DUT.

For instance, let the DUT have one output and be described by the equation

$$y(\tau) = f(y(\tau - 1),\, y(\tau - 2),\dots,\, y(\tau - m)),$$

where every argument can be absent.

The structure of ring testing system for such a DUT is shown in Figure 7.12A. It is composed of a correction unit and an XOR gate. The correction unit, in its turn (as will be shown further) is composed of an m-digit shift register and a certain combinational circuit. In such a system, the correction unit is devised to generate the required T period and to record the test results, using the m-digit shift register. The feedback function is given as

$$\varphi(\tau) = \sum_{i=1}^{m} a_i\, \varphi(\tau - i),$$

whereas the T period is identified with the irreducible polynomial exponent

$$g(z) = \sum_{i=1}^{m} a_i z^{m-i} \quad (a_0 = a_m = 1).$$

Example 5 [3]

Let the DUT be described by the ratio

$$y(\tau) = y(\tau - 2)\, y(\tau - 3)$$

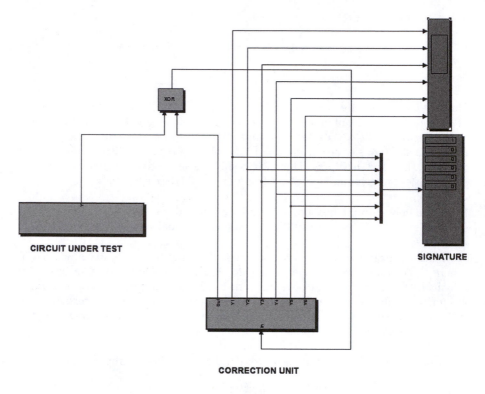

FIGURE 7.12A
Ring testing system structure for the DUT represented by an output-independent FSM.

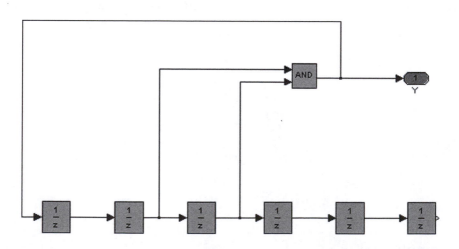

FIGURE 7.12B
Structure of the DUT represented by an output-independent FSM.

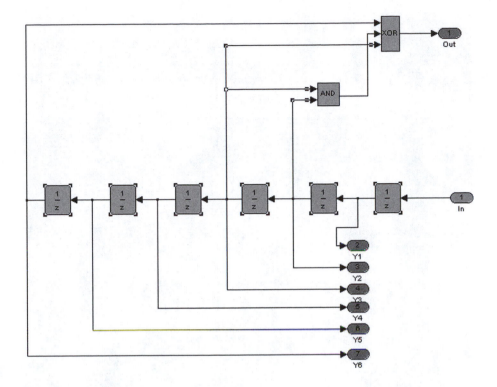

FIGURE 7.12C
Structure of the correction unit.

(the DUT model of Figure 7.12B is constructed from this expression). The ring testing system of T = 9 period must be constructed.

In this case (as before) the feedback polynomial is given as

$$g_9(z) = z^6 + z^3 + 1.$$

Hence, it follows that

$$\varphi(\tau) = y(\tau - 3) \oplus y(\tau - 6),$$

and the correction unit function is

$$F(\tau) = \varphi(\tau) \oplus y(\tau) = y(\tau - 3) \oplus y(\tau - 6) \oplus y(\tau - 2) \, y(\tau - 3).$$

The correction unit structure corresponding to the function is shown in Figure 7.12C. It contains the six-digit shift register (m = 6) and the AND and XOR gates.

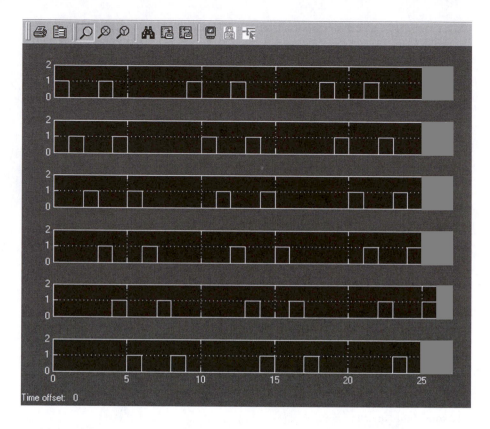

FIGURE 7.12D
Test results. Strict periodicity (T = 9) testifies to the absence of faults.

By system setting into its initial condition, the DUT state and the correction unit $y_1(0)$, $y_2(0)$,..., $y_m(0)$ state should be made compatible. If faults are absent, the system performance is strictly periodical, and, with m = 6, is described by

$$\mathbf{Y}_6\,(\tau + 1) = \mathbf{H}\;\mathbf{Y}_6\,(\tau) = \mathbf{H}^{\tau+1}\,\mathbf{Y}_6\,(0)\;(\tau = 0, 1, 2...),$$

where: $\mathbf{Y}_6\,(\tau) = [y_1(\tau),..., y_m(\tau)]$ is a column vector of the register outputs, $\mathbf{H}$ is the accompanying matrix of the irreducible g(z) polynomial, and deg g = m.

The fault-free DUT is determined by observation of shift register outputs in the T tact, for which the following equality should be held (Figure 7.12D):

$$\mathbf{Y}_6\,(T) = \mathbf{Y}_6\,(0).$$

Linear discrete devices used for the construction of numerical filters, generators, coders, and decoders are the most convenient ring DUTs. The ring testing system has the lowest hardware complexity for linear discrete devices, as the system correction unit does not possess the DUT linearization provision function. The correction unit ensures only the desired system performance period. In some cases the correction unit is of no need at all.

The combinational linear discrete unit with n inputs and m outputs is described by

$$Y = AX, \tag{7.20}$$

where $X = (x_1,..., x_n)$ is the input column vector, $Y = (y_1,..., y_m)$ is the output column vector, and A is a matrix over GF(2) with $m \times n$ dimensionality.

For a linear discrete device, the

$$\Phi(x_1,..., x_n) = y_1 \oplus \cdots \oplus y_m$$

function can be recorded as

$$\Phi(x_1,..., x_n) = b_0 \oplus b_1 x_1 \oplus \cdots \oplus b_n y_n,$$

where $b_i = 0$ or 1. The ring testing system T period is defined from the feedback function

$$\phi(x_1,..., x_n) = a_1 x_1 \oplus \cdots \oplus a_n y_n,$$

which has the correspondence in the g(z) polynomial over GF(2), belonging to the same T exponent. In this process the correction unit is synthesized according to

$$F(x_1,..., x_n) = \phi(x_1,..., x_n) \oplus \Phi(x_1,..., x_n).$$

$F(x_1,..., x_n)$ is a linear function

$$F(x_1,..., x_n) = c_0 \oplus c_1 x_1 \oplus \cdots \oplus c_n y_n, \ (c_i = 0 \text{ or } 1)$$

because the class of linear Boolean functions is closed relative to the addition by mod2. Hence, it follows that the correction unit needs for its realization less than n connections of shift register outputs, the XOR gates inputs, and the c_0 constant.

Example 6 [3]

Let the circuit under test (CUT) be a code-former whose diagram is shown in Figure 7.13A. The code is formed according to Equation 7.20, where matrix **A** takes the form of

$$A = \begin{Vmatrix} 1111111110000000 \\ 0101010111010100 \\ 0011001100110011 \\ 0000111100001111 \\ 1000000011111111 \\ 1110100101101001 \end{Vmatrix}$$

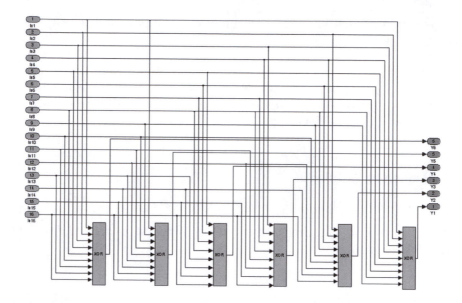

FIGURE 7.13A
Structure of the CUT-representing code-former.

It follows from the Equation 7.20 relation that

$$y_1 = x_1 \oplus x_2 \oplus x_3 \oplus x_4 \oplus x_5 \oplus x_6 \oplus x_7 \oplus x_8 \oplus x_9;$$

$$y_2 = x_2 \oplus x_4 \oplus x_6 \oplus x_8 \oplus x_9 \oplus x_{10} \oplus x_{12} \oplus x_{14} \oplus x_{16};$$

$$y_3 = x_3 \oplus x_4 \oplus x_7 \oplus x_8 \oplus x_{11} \oplus x_{12} \oplus x_{15} \oplus x_{16};$$

$$y_4 = x_5 \oplus x_6 \oplus x_7 \oplus x_8 \oplus x_{13} \oplus x_{14} \oplus x_{15} \oplus x_{16};$$

$$y_5 = x_1 \oplus x_9 \oplus x_{10} \oplus x_{11} \oplus x_{12} \oplus x_{13} \oplus x_{14} \oplus x_{15} \oplus x_{16};$$

$$y_6 = x_1 \oplus x_2 \oplus x_3 \oplus x_5 \oplus x_8 \oplus x_{10} \oplus x_{11} \oplus x_{13} \oplus x_{16}.$$

As a result we have the function of

$$\Phi(x_1,\ldots, x_n) = y_1 \oplus \cdots \oplus y_m = x_1 \oplus x_2 \oplus \cdots \oplus x_{16}.$$

If we want to do without a correction unit, the system period will be defined by the

$$g(z) = z^{16} + z^{15} + \cdots + z + 1$$

polynomial over GF(2) that generates the T = 16 period.

The structure of ring testing system without a correction unit is shown in Figure 7.13B. Figure 7.13C shows the shift register structure for the system. The results of system simulation with a fault-free DUT are shown in Figure 7.13D.

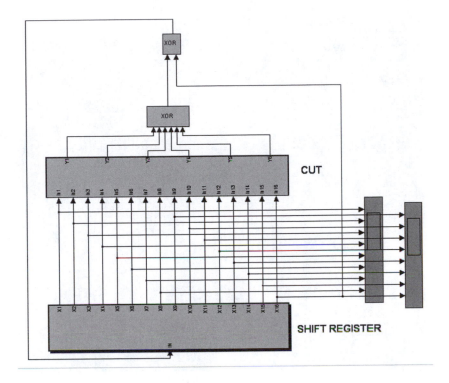

FIGURE 7.13B
Structure of the correction-unit-free ring testing system.

7.3.2 Diagnostic Testing

The diagnostic device structure is shown in Figure 7.14A. It is composed of the test vector source, the reference object (FAULT-FREE CIRCUIT), the DUT (FAULT CIRCUIT), the comparator (COMPARATOR), the diagnostic machine (Chart), the decoder (DECODER), and the result-indicating blocks.

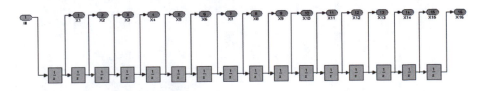

FIGURE 7.13C
Structure of the system shift register.

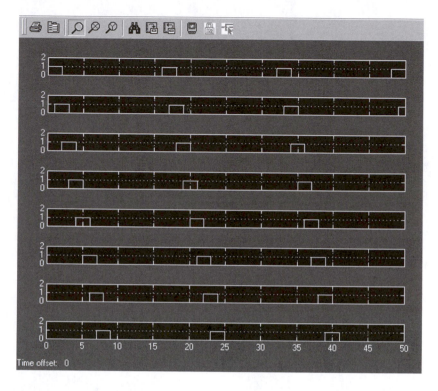

FIGURE 7.13D

Test results. Strict periodicity (T = 16) testifies to the absence of faults in the code-former.

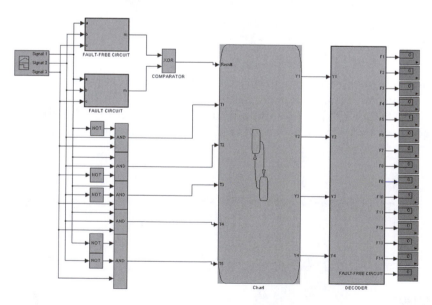

FIGURE 7.14A

The diagnostic scheme structure.

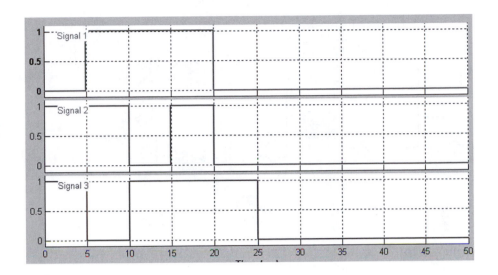

FIGURE 7.14B
Time diagrams of diagnostic tests.

We will give an example taken from [6]. The fault-free circuit represents a circuit containing 13 nodes (Figure 7.14C): a, b, c, d, e, f, g, h, i, j, k, l, and m. Thus, a fault circuit can have up to 26 faults of the single stuck fault (SSF) type. During fault detection five test vectors are used (Table 7.2, Figure 7.14B).

We can make a table that will tabulate the node states of the circuit exposed to the above test vectors and contain, at a certain moment, 1 of 26 possible (potential) faults of the SSF type (Table 7.3). Each row of Table 7.3 corresponds to a definite test vector, while each column corresponds to a definite fault (for instance, a_0 denotes a fault of the stuck-at-0 type in the **a** node). The asterisk denotes the difference between the reference and the actual states of the circuit node. Instead of 26 columns for 26 possible faults of the

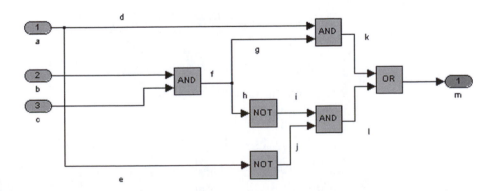

FIGURE 7.14C
The reference object structure.

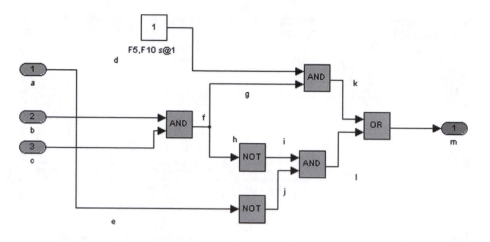

FIGURE 7.14D
The DUT structure (containing the F5, F10 faults of the stuck-at-1 type).

SSF type, the table has only 14 columns, which means the circuit diagnostics are not full: each fault cannot be detected with the access granted to the circuit a, b, c inputs only. Therefore, it sounds logical to subdivide the integrity of all possible faults into 14 equivalent classes of F_i (i = 1,2,..., 14) and to conduct the detection of these equivalent faults:

$$F1: \{a_0\}$$

$$F2: \{a_1\}$$

$$F3: \{b_1\}$$

$$F4: \{c_1\}$$

$$F5: \{d_1\}$$

$$F6: \{b_0, c_0, f_0\} \Rightarrow \{f_0\}$$

$$F7: \{f_1\}$$

$$F8: \{g_1\}$$

$$F9: \{e_1, h_1, i_0, j_0, l_0\} \Rightarrow \{i_0\}$$

$$F10: \{i_1, h_0\} \Rightarrow \{i_1\}$$

$$F11: \{j_1, e_0\} \Rightarrow \{j_1\}$$

$$F12: \{g_0, d_0, k_0\} \Rightarrow \{k_0\}$$

$$F13: \{k_1, l_1, m_1\} \Rightarrow \{k_1\}$$

$$F14: \{m_0\}$$

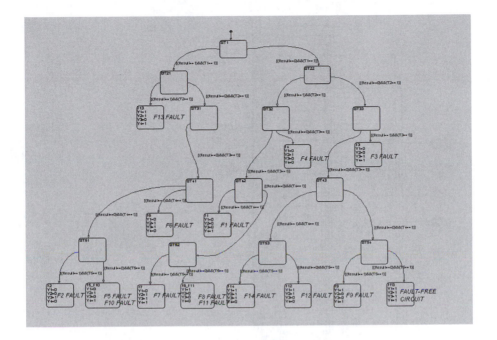

FIGURE 7.14E
Fault tree for the Figure 7.14C circuit.

These equivalent faults make up the fault dictionary for a given case. The fault tree is designed in conformity with the dictionary and is realized using the Stateflow Chart Block (Figure 7.14E). Descending the tree to a leaf corresponding to one or several equivalent faults, F_i, is the detection process. During the process, each decision is based on the result of an elementary test t_i, consisting of one i-th test vector (Fail, Pass). Figure 7.14F shows the structure of a decoder from the Figure 7.14A circuit.

7.3.3 JTAG Interface Model

The joint test action group (JTAG) interface is a totality of facilities and operations permitting the user to test VLSIs without physical access to each of their outputs. The testing according to the IEEE Std 1149.1 standard is called boundary scan testing (BST). Such testing is practicable only for chips with an inside set of special elements, that is, the boundary scan cells (BSCs) and their operation control schemes. Later on, the JTAG interface functions were expanded and used extensively in the configurations of programmable logical devices [1, 2, 7].

When the JTAG interface was being developed, the key requirement was the minimization of the VLSI contacts used for information exchange during testing. There are four (sometimes five) contacts, and their integrity forms the test access port (TAP). The contacts destinations are:

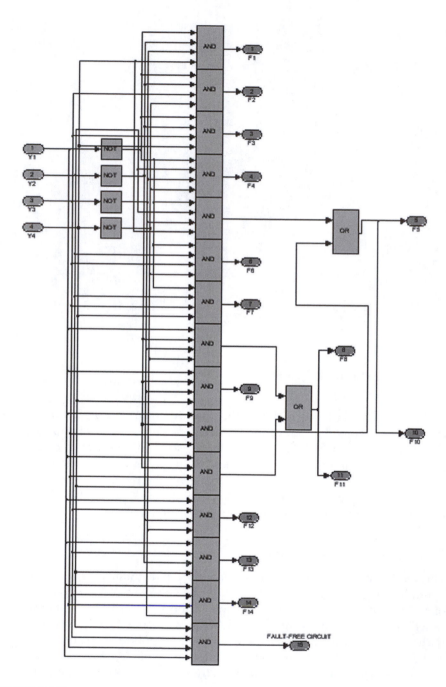

FIGURE 7.14F
Structure of a decoder from the Figure 7.14A circuit.

TABLE 7.2

Test Vectors for Fault Detection

Test Vectors	a	b	c
t_1	0	1	1
t_2	1	1	0
t_3	1	0	1
t_4	1	1	1
t_5	0	0	1

TCK: For the data transfer and instruction synchronization signal

TMS: For the transmission mode selection

TDI: For data and instruction input

TDO: For data, instruction, or state output

TRST (if employed): For TAP controller return to the initial state

Such a small number of contacts is sufficient because of successive transfer of instructions and data.

7.3.3.1 BSC Model

The key concept of BST is shown in Figure 7.15. The BSC cells are located between each chip outer output and the DUT-generating crystal circuits and can operate in different conditions. In working conditions they only allow the signals pass through from left to right and do not change the device operation. Input signals pass through the BSC cells directly to the corresponding points of the crystal basic circuits. In doing this, the conventional logical-type outputs are equipped with one BSC cell, the third-state outputs need two BSC cells (the second one is engaged in generating the buffer control signal), and the bidirected outputs need three BSC cells. However, the so-called passive testing is practicable here as well. The BSC cells can be joined in series to form shift registers, where the values of transmitted signals are recorded and later read by the tester.

TABLE 7.3

Potential Faults of the SSF Type

	a_0	a_1	b_1	c_1	d_1	f_0	f_1	g_1	i_0	i_1	j_1	k_0	K_1	m_0	Fault Free Circuit
t_1	0	1*	0	0	1*	1*	0	0	0	1*	0	0	1*	0	0
t_2	1*	0	0	1*	0	0	1*	1*	0	0	1*	0	1*	0	0
t_3	1*	0	1*	0	0	0	1*	1*	0	0	1*	0	1*	0	0
t_4	0*	1	1	1	1	0*	1	1	1	1	1	0*	1	0*	1
t_5	1	0*	0*	1	1	1	0*	1	0*	1	1	1	1	0*	1

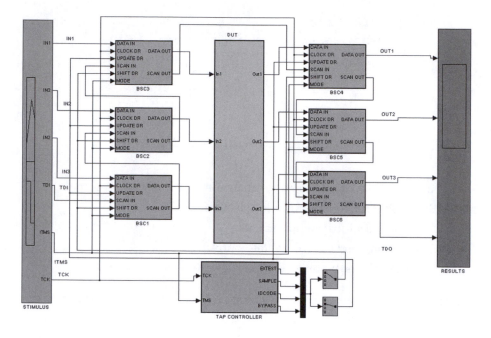

FIGURE 7.15
JTAG interface model.

In active testing conditions signal transmission through the cells ceases, and the test code can be successively fed from the tester to the shift register and then rewritten to the static lock register and fed to the input points of the crystal basic circuit. Both registers (shift static) are created from the cell resources. The result generated by the basic circuit can be transferred to the BSC output cells and then withdrawn out of the VLSI successively, in the shift register mode, for comparing with the expected reference result (the result analysis as well as the input data preparation for testing is made by an external tester).

The BSC cell circuit (Figure 7.16) contains two multiplexors and two D-type flip-flops (DFFs). As dictated by the address input MODE of output multiplexor, the cell can either freely pass the signal from input to output or transmit the state of the second DFF, creating the static register DR (data register) discharge, to the output. The input multiplexor SHIFT DR address signal controls the input signal (from chip logical inputs) feed or the signal from the BSC's previous cell to the first DFF. Thus, following the signal feed from the multiplexor's lower input (SCAN IN) to the integrity of first DFFs, they form the shift register, and following the signal feed from multiplexor's upper input (DATA IN) to the integrity of first DFFs, the DFFs are loaded in parallel by the data from VLSI input contacts. When the previous cell signal (SCAN IN) is fed through the input multiplexor, the CLOCK DR signal makes

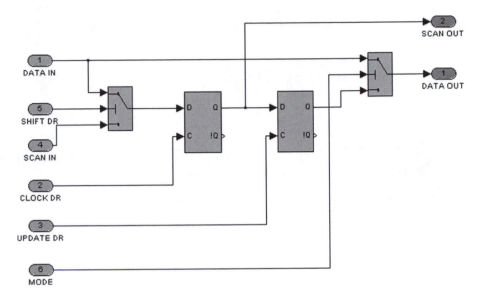

FIGURE 7.16
Boundary scan cell (BSC) model.

a one-digit shift in the register formed by and in-series connection of first DFFs. Parallel loading of the same DFFs is also set by the CLOCK DR signal.

Following the UPDATE DR synchronous signal, current content of the register composed of the first DFF chain is rerecorded to the static register composed of the second DFFs. The shifts in the first-DFF register will not affect the second-DFF register content.

7.3.3.2 BSC Chain Model

If a DUT has several VLSIs with a JTAG interface, they can be pooled into the JTAG chain (Figure 7.17). When any operation is executed, instructions and data are transmitted along the chain consecutively. Using only the test mode select (TMS) and test clock (TCK) contacts, the JTAG chain control device, incorporated into the tester, can set the TAP controller machines of all chain VLSIs into any desired state (starting state, instruction or data loading to the registers, reading of register data). The totality of registers of all chain VLSIs seems to be a single register on the route from the tester output contact test data in (TDI) to its input TDI. Therefore, while adjusting one of the chain VLSIs, a bit sequence, with a length corresponding to the entire chain, should be composed and introduced into the chain (it is applied both to the instruction register chain and to the data register chain). When the data and instruction transmission modes are actuated, each TCK pulse shifts the register chain code by 1 digit.

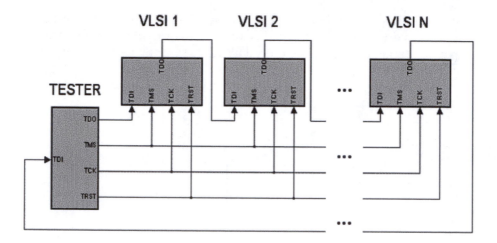

FIGURE 7.17
BSC chain model.

The boundary scan is made in the following sequence:

1. Instruction loading
2. Data loading
3. Instruction execution
4. Result reading

Functional potentialities of BSC cells permit the establishment of various boundary scan modes:

1. The VLSI self-test or recording or reading of intracircuit memories. In this condition, the VLSI outputs are isolated from its inner circuits, to which the data from the second DFF cells of the BSC JTAG chain are fed. Subsequent operations are dictated by an input instruction. The resulting information (response to the input data fed) can be recorded in the first DFFs of the BSC cells and transferred to the tester for analysis, which is exactly required for VLSI self-testing or for recording or reading of intracircuit memories.

2. Testing of VLSI interconnections. In this mode, the VLSI outer contacts are also disconnected from its inner circuits. The data are loaded to the JTAG chain output cells of one VLSI and are further transferred to the JTAG chain input cells of the second VLSI. If the VLSI interconnections are fault-free, the data should coincide, and the tester checks it.

3. Testing of VLSI standard operation. Here, the crystal inner circuits are connected to VLSI outer contacts, thus corresponding to its performance conditions. At a preset moment, the BSC cells record the

conditions in all contacts. Transfer of the data obtained to the tester allows estimation of the validity of VLSI performance. In this process, the circuit inner signals become known even without any physical access to the circuit's contacts.

4. The mixed conditions, when some VLSIs are in standard conditions, whereas others are in test conditions.

7.3.3.3 TAP Controller Model

The BSC device is composed of the parallel-connected instruction register (IR), data register (DR), and BYPASS register, as well as output multiplexor and the TAP controller. The DR (scan register) is produced by an in-series connection of the BSC cell's first DFFs and accepts or yields data when any instructions are executed in the JTAG chain. The register parallel connection means the existence of common in-series input and output, TDI, and TDO. A single-digit (passing) BYPASS register is used in data loading and discharge conditions as a bypass for the shifted multidigit data, not pertinent to said VLSI. Transmission of data from TDI input to TDO output via the BYPASS single-digit register reduces the chain length and accelerates testing processes.

The TAP controller (Figure 7.18 and Figure 7.19) receives the TCK clock signals and interprets the TMS input instructions by choosing one or another register for recording or reading. The instructions recorded in lock registers are decoded and generate the signals that choose the boundary scan conditions or the data registers that are connected to TDO output via the multiplexor.

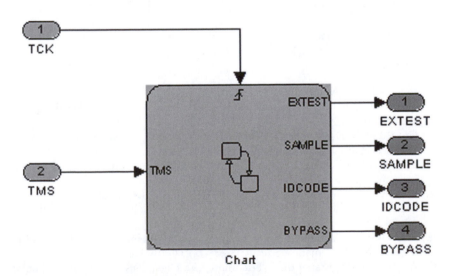

FIGURE 7.18
TAP controller model.

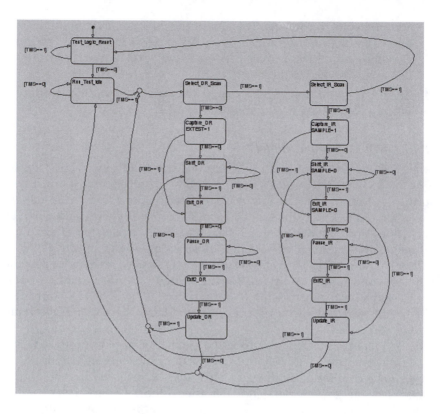

FIGURE 7.19
TAP controller model (state diagram).

References

1. Bushnell M.L., Agrawal V.D. Essentials of Electronic Testing for Digital, Memory & Mixed-Signal VLSI Circuits. Kluwer Academic Publishers, Dordrecht, The Netherlands, 2004.

2. Crouch A.L. Design-for-Test for Digital IC's and Embedded Core Systems. Prentice Hall PTR, Reading, MA, 1999.

3. Litikov I.P. Ring Testing for Digital Units (in Russian). Energoatomizdat, Moscow, 1990.

4. Rabaey J.M., Chandrakasan A., Nicolic B. Digital Integrated Circuits: A Design Perspective (2nd ed.). Pearson Education International, Upper Saddle River, NJ, 2003.

5. Zhegalkin I.I. Arithmetic for symbolic logic. Mathematical Collected Volume of Moscow Mathematical Society, 36(3-4), 1929 (in Russian).

6. Abramovici M., Breuer M.A., Friedman A.D. *Digital Systems Testing and Testable Design*. IEEE Press, New York, 1995.

7. Parker K.P. *The Boundary-Scan Handbook* (3rd ed.). Kluwer Academic Publishers, Dordrecht, The Netherlands, 2003.

Index